Apollo in Perspective

Spaceflight Then and Now

Apollo in Perspective

Spaceflight Then and Now

Jonathan Allday

The King's School, Canterbury

Institute of Physics Publishing
Bristol and Philadelphia

British Library Cataloguing-in-Publication Data

A catalogue record for this book is available from the British Library.

ISBN 0 7503 0644 0 (hbk)
0 7503 0645 9 (pbk)

Library of Congress Cataloging-in-Publication Data are available

Production Editor: Al Troyano
Production Control: Sarah Plenty and Jenny Troyano
Commissioning Editor: Jim Revill
Editorial Assistant: Victoria Le Billon
Cover Design: Jeremy Stephens
Marketing Executive: Colin Fenton

Published by Institute of Physics Publishing, wholly owned by The Institute of Physics, London

Institute of Physics Publishing, Dirac House, Temple Back, Bristol BS1 6BE, UK

US Office: Institute of Physics Publishing, The Public Ledger Building, Suite 1035, 150 South Independence Mall West, Philadelphia, PA 19106, USA

Typeset in TEX using the IOP Bookmaker Macros
Printed in the UK by J W Arrowsmith Ltd, Bristol

‘Men who have worked together to reach the stars are not likely to descend together into the depths of war and desolation.’

Lyndon B Johnson (1958)

‘For a successful technology, reality must take precedence over public relations, for Nature cannot be fooled.’

Richard Feynman
Appendix to the Presidential Report on the Challenger Space Shuttle Disaster

Contents

Introduction

To the Moon, by washing machine

At one of the pivotal moments in the film *Apollo 13* Jim Lovell's wife is explaining to Lovell's mother that the mission is in jeopardy and the astronauts may not make it back to Earth alive. Lovell's mother is unfazed by this: 'if they could get a washing machine to fly, my Jimmy could land it'.

I was reminded of this exchange recently while talking to an intelligent 13-year-old student about the Apollo missions. He was quite adamant that the journeys had never taken place and that the whole thing had been an elaborate hoax. Putting on my best Fox Mulder voice, I asked what could have possibly made him believe such an extraordinary thing.

He thought for a moment then said, 'Well, did you know that the computers in those days were not as good as the ones you find now in washing machines?'

'That's true,' I replied, 'they were not as good as the ones you find running car engines either.'

'Well, there you are,' he said, 'you couldn't possibly get to the Moon using such computers.'

I looked at him rather blankly for a few moments, until I realized that he was *serious*. I was about to point out that people like Jim Lovell could probably fly washing machines, but I lost my chance as the conversation had moved on to a discussion about the lighting angles of certain photographs taken on the Moon's surface and how the reflection in one of the astronaut's faceplates had been wrong.

While musing over this conversation later I realized that this young man, brought up in a world of conspiracy theories and *X-Files*, was as excited by believing that there was a huge government lie being uncovered as I had been following what really happened 30 years ago.

On 20 July 1969 mankind's future changed utterly. I was awake at 4.00 am and watched Neil Armstrong step off the footpad of the Apollo 11 lunar module onto the surface of the Moon. In one sense I had no idea of the significance of the event. The politics, the business competition and the fears of the people involved (for example, they were not sure that the surface of the Moon was strong enough to support the lunar module's weight) were unknown to me. On the other hand, I was fully aware of one of its most important aspects—it was exciting. I was transfixed.

Space had been part of my life ever since I was old enough to start reading Captain W E Johns' books (not Biggles, the science fiction ones that he is not as well known for). I had read *2001* and had been totally baffled by the film. Now I was watching the whole thing come alive in front of me. Anyone with a lively imagination would have been captivated.

Yet now it is 30 years on. It is all part of history. Today's young people are mostly unaware of the events that took place in 1969. They know that men have been to the Moon, but they do not know why or, more importantly, how. The excitement has gone and will probably not be recaptured until we finally send a manned expedition to Mars. The film and photographs brought back from the Moon recapture the sense of the times for those who saw it live, but do little for the modern watcher. They are far more likely to laugh at the strange clothes worn by the sixties occupants of mission control (and they are quite funny).

Of course, this is perfectly understandable. It is now a different world and the young people in it have different expectations. However, I maintain that the Apollo missions are a vital part of human history and that everyone should have some appreciation of how they were done.

Why?

Partly because of the technology involved. These days so little seems to be beyond us that it is important to try to recapture the sense of pushing technology to the edge—how else will we motivate students to consider engineering and the technological sciences?

Partly as an excuse for trying to explain a little bit of important physics. As far as I am aware, of all the books that have been written on the Apollo programme, none has tried to explain in simple terms the physics and engineering involved—which is the primary aim of this book.

However, this book is about more than just the Apollo programme. It is about Apollo judged from our perspective at the end of the twentieth century. From this vantage point we can also chart the progress in space science since that time—progress that seems to have been halting and hesitant since the heady days of Apollo when they seriously thought that they could have a man on Mars by the 1980s. However, progress has been made and I shall argue here that another explosion of interest and activity in space is probably about to happen in the early part of the next century. We shall look at where that may take us.

Finally, looking back at Apollo is timely because so few people these days seem to think that it is important. It is hard to argue against those that say that the money would have been better spent on feeding the poor etc. We can say that the money would not have been available for that purpose—a sad human truth—but that hardly seems sufficient. The best defence I have heard of the space programme comes from J M Straczinsky as spoken by Commander Jeffrey Sinclair in the science fiction series *Babylon 5*. Sinclair is being interviewed by a reporter about the expense of maintaining a space station—the standard question is asked—can the expense be justified?

> 'Ask ten different scientists about the environment, population control, genetics . . . and you'll get ten different answers. But there's one thing every scientist on the planet agrees on. Whether it happens in a hundred years, a thousand years or a million years eventually our Sun will go cold and go out. When that happens it won't just take us—it'll take Marilyn Monroe, Lau Tzu, Einstein and Buddy Holly and Aristophanies—all of this, all of this was for nothing. Unless we go to the stars.'

On 20 July 1969 mankind took the first vital step to securing its indefinite future. For ever more that date will have significance.

I wonder how future historians will judge the 30 years that followed?

Acknowledgments

As always with a book like this, there is a long list of people that I need to thank, as without them it would not have been written. It is quite possible that I have missed somebody, in which case I apologise to them—they can write to me and I will correct it for the second edition! (HINT).

My parents—as always. Both books that I have written have been about something that I was passionately interested in. They are probably as puzzled as I am about where particle physics came from, but I know that my interest in space (and Apollo especially) comes from their interest and encouragement. I was a captivated 9 year old in front of a television taping the conversations between ground and the astronauts. Without their selfless support in all manner of ways this book would not have happened. They also did a great deal of useful research for me on the Internet.

Carolyn—who is remarkably tolerant, both of her overworking husband and our three energetic sons. Next year will be the year off that I promised. Probably.

Ben, Josh and Toby—I hope that they will get the chance to sit with their dad and watch the first humans walk on Mars.

John and Margaret Gearey—that desk in the study has got something about it!

Jim Revill at Insitute of Physics Publishing—for believing that lightning does occasionally strike twice in the same place in quick succession.

Al Troyano and Victoria Le Billon for great patience and no little skill in producing this book.

The reviewers, Robert Strawson (Abingdon School) and Hal Heaton (Johns Hopkins University), for their sympathetic and detailed review of the manuscript. Their efforts have made this a better book than it otherwise would have been.

NASA, for making this all possible in the first place and then for placing so much information in the public domain through their excellent websites. Many of the images in this book have been accessed from NASA websites or have been supplied directly by NASA for which I am very grateful. The cover image is taken from an Apollo 17 photograph of Mission Commander Gene Cernan on the lunar surface and is used with the kind permission of NASA.

Larry A Feliu of the NorthropGrumman History Centre for timely information on the lunar module.

Kip Teague for help with scans of various pictures posted on the net.

Dick Churcher—for helping out with atmospheres and lunar geology—a good man to have around whenever the balloon goes up.

Ian Longfellow—for lots of conversations that made me think, and some help with computers.

Also Apple computers—the last best hope.

Jonathan Allday
(jonathan@jacant.demon.co.uk)
Canterbury, August 1999

A note about units

I have stuck with the standard SI units in this book (i.e. kilograms, metres and newtons etc). Conversion factors can be found in the glossary at the back of the book.

Colour photographs

Many of the photographs in this book can be viewed in colour at the following website

http://bookmark.iop.org/bookpge.htm?ID=&book=797p

Chapter 1

Apollo in outline

1.1 The mission

The plan to land a man on the Moon was deceptively simple.

A Saturn V booster rocket lifted a three-man crew and their equipment from the surface of Earth. The booster was built in three sections (called stages) which had independent propellant tanks and rocket engines. The advantage of such an approach is that after the propellant tanks in one stage have been emptied they can be ejected, along with the rest of the stage. This saves carrying their considerable mass any higher. The compensation in terms of the total amount of propellant required outweighs the additional mass of the extra engines needed.

Each stage burnt its engines for a set period of time, after which explosive charges fired separating the empty casing from the rest of the rocket. Then the next stage's engines automatically ignited pushing the rocket higher.

By the time that the third stage's engine had shut down the Apollo astronauts were in orbit. After a period of time circling the Earth while the crew and ground support teams thoroughly checked out the equipment, the third-stage engine fired again injecting the craft into a much larger orbit that would take it out towards the Moon.

Soon after *trans-lunar injection*, the *command module* (in which the crew lived), attached to the *service module* (which contained propellant, air and electrical power generating equipment), separated from the third stage of the booster. The command module pilot moved the combination (known as the CSM) a safe distance away, turned it round and brought it back to the third stage, nestling inside which was the *lunar excursion module* (LM). This was the craft that had been designed to land on the surface of the Moon. The command module

Figure 1.1 The launch of the Saturn V booster that carried the Apollo 11 spacecraft.

docked with the LM and slowly extracted it from its housing[1]. Finally the joined craft backed away from the third stage and proceeded to the Moon in tandem. All this took place while both craft were moving through space at a speed of about 11 km per second. Of course, this is not as dramatic as it sounds as their relative speeds were not that great!

About three days later the command, service and lunar module combination arrived at the Moon. For the first landing the trajectory of the craft had been calculated to swing it round behind the Moon and return to Earth. This *free return trajectory* was designed to bring the astronauts home automatically in case of a problem. (Later missions deviated from this trajectory in order to increase the number of regions on the Moon that were accessible for a landing.) In order to enter orbit round the Moon the spacecraft had to slow down by firing the service module engine as the astronauts were passing behind the Moon. Unfortunately, this meant that one of the most critical manoeuvres in the mission was carried out while the astronauts were out of contact with Earth. Mission control would not know if the 'burn' had been successful until

Figure 1.2 The Apollo 15 command and service module combination in lunar orbit. The command module is the blunt cone at the right-hand end of the cylindrical service module which has a panel open so that a camera can take pictures of the lunar surface.

the radio link was established again as the craft came round the other edge of the Moon. The precise time delay between the moment at which contact was lost, as Apollo passed behind the Moon, and when it was re-established, as it emerged round the other side, had been carefully calculated on the basis of a successful burn. Mission control could tell that the astronauts were in orbit from this timing.

Once safely in orbit, the mission commander and lunar module pilot crawled into the LM and started to check out its systems. All being well, the LM detached, leaving the command module pilot alone. The LM was then flown in front of the CSM while the command module pilot gave it a visual check. Finally, the LM's descent engine was fired to move onto a new orbit. This took them lower down so that they could scout out the landing area and get used to the sequence of lunar features that would appear on their landing trajectory. After a couple of orbits doing this, the descent engine was fired once again to put the LM on another new trajectory—the *powered descent*. Under engine power the two astronauts flew the LM down to a pre-defined landing area. However, as the best photographs of the Moon could not detail all the features, the final choice of landing spot was left open to the pilot.

Figure 1.3 The Apollo 16 lunar module on the surface of the Moon.

Neil Armstrong and Buzz Aldrin carried out the first landing—and it was far from routine. From the start there were communications difficulties between ground control and the lunar module. At times messages had to be relayed through the command module. Then as the powered descent was taking place problems with the programmed trajectory began to emerge. Although the landing trajectory had been worked out in advance, the mission planners were not able to take everything into account, as some aspects of the information that they needed were unknown. For example, local regions of Moon rock with a higher than average density would pull on the LM with a greater than average gravitational force, deflecting the course. There was no way of knowing about these mass concentrations (mascons) until they flew over one.

For whatever reason, the computer on the Apollo 11 LM got things slightly wrong, and as they were flying over the lunar surface Armstrong noticed that all the landmarks that he had been trained to look out for were appearing about 2 to 3 seconds before he expected them. At the speed that they were travelling that translated into overshooting the landing spot by 2–3 miles. It turned out that the computer was aiming for a large crater full of boulders, some of which were the size of cars. Armstrong took over flying the LM and spent tense moments hovering about looking for a safe place to land. To cap it all the LM computer was periodically throwing up ‘program alarms’—the equivalent of the error messages that we get on today’s PCs. The computer was having too much to do in the regular cycle of information that it had to check through and was complaining of being overworked. Fortunately the programmers who had written the code for the machine were on hand in mission control and they confirmed that the computer was just complaining and the errors did not represent a serious problem. Armstrong, meanwhile, was still searching for a safe landing spot and the tension at mission control was climbing. The mission plan called for an upper limit on the amount of propellant that could be used during the descent. If anything went wrong the abort plan was to fire the descent engine to boost them away from the surface and then to separate the ascent stage and fire its engines for the return to orbit. With the LM 6 m above the ground, Houston called out (with some urgency) to the astronauts that they had 30 seconds of propellant left before the abort margin was reached. When the descent engine finally shut down as they settled onto the surface, they had about 45 seconds of propellant remaining.[2]

Apollo 11 had landed, however nobody knew quite where! The geologists in mission control were not best pleased as they had planned some serious scientific work at the landing site. In orbit, Mike Collins, the command module pilot, was sent information from Earth to try to look for the LM through his telescope! Although he continued to try until the explorers returned to orbit, he never caught a glimpse of them.

Although the problems surrounding the Apollo 11 landing were not widely known outside NASA, pride dictated that the next landing would be a little more ideal. Not just that, the geologists planning for the later missions wanted to be sure that the astronauts would be collecting rocks from the right area. Some sources of error in the original landing plan were sorted out, but the key to a precision landing turned out to be continual monitoring of the radio transmissions from the LM. As the LM moved across the lunar surface its motion altered the frequency of the radio signal in a similar way to that in which the motion of an ambulance or fire engine will alter the sound of the siren. Tracking this shift in frequency and comparing it to the expected shift calculated from the trajectory that the LM should be following enabled mission control to keep a continual check on the actual motion of the craft. Any errors could then be relayed to the LM so that the astronauts could compensate.

It was decided that Apollo 12 would attempt to land within walking distance of the Surveyor 3 probe which had soft-landed on the Moon on 20 April 1967. The idea was to inspect what long-term exposure to space had done to the probe. The site had also been listed as being of geological interest, so the two objectives dovetailed nicely. It would also not hurt to show what could be done in terms of a pinpoint landing. It was a risky strategy. Failure would be very public. However, the fear of failure is a great motivator and NASA was used to taking risks.

The landing was made using the new guidance system, which only needed a small manual correction in the last phase of the landing. The amount of dust thrown up by the descent engines was far greater than had been the case during the Apollo 11 landing, so Pete Conrad, who was piloting the craft, could see little through the windows from about 30 m above the surface. Touchdown

Figure 1.4 One of the Apollo 12 astronauts examines the Surveyor 3 probe with the lunar module in the background.

was very smooth, and within moments of placing his foot on the surface, Pete Conrad could see the probe on the side of a crater. Although Surveyor 3 was too small to be seen by the astronauts during the landing, they touched down within 160 m of the probe. Considering that the astronauts had travelled about 1.2 million kilometres on their flight from Earth, this represents extraordinary precision.

After Apollo 12 came the famous Apollo 13 mission in which a serious incident involving the service module crippled the spacecraft and placed the astronauts' lives in jeopardy. Thankfully some brilliant improvization on the part of mission control and the resilience of the astronauts brought the flight to a safe conclusion. Ironically the Apollo 13 mission did serve to re-awaken interest in the Apollo programme at a time when the public was beginning to feel that flights to the Moon were becoming routine. In fact, the serious exploration of the surface was just shifting into a higher gear.

The amount of time spent on the surface increased as the missions went on and the planners gained more confidence in the astronauts' ability to work on the Moon. The Apollo 11 crew had two and a half hours of frantic flag planting

Figure 1.5 The Apollo 14 command module just prior to splash down.

and rock gathering, whereas on the final mission, Apollo 17, the crew explored for a total of 22 hours using a small car to drive round, covering 35 km in total. All the astronauts were trained in geology, but it was not until the Apollo 17 mission that a geologist, who had been trained as an astronaut, was sent to the Moon.

To start the return journey the LM's ascent engine fired, separating the top half, which comprised the cabin, from the bottom of the craft containing the descent engine and the landing legs. The ascent engine carried the astronauts back into orbit and a rendezvous with the command module pilot still orbiting in the CSM. After docking the astronauts transferred their samples, cameras and film back to the CSM and jettisoned the now redundant ascent stage of the LM.

Once again, behind the Moon, the CSM engine fired to break the craft out of orbit onto a trajectory to bring it back to Earth. The return journey would take another three days.

Shortly before arriving at Earth the astronauts jettisoned the service module and turned the command module round so that its underside was facing the Earth. Strapped into couches they entered the Earth's atmosphere backs first. Beneath them the underside of the command module comprised a specially designed shield that protected them from the heat of re-entry.

Figure 1.6 The Apollo 11 isolation chamber is unloaded from the aircraft carrier USS Hornet (with the astronauts inside!). The chamber was taken to the NASA lunar receiving lab and the crew was released on 10 August.

The craft hit the atmosphere at 40 000 km per hour.

Rapidly the temperature climbed to 5000 °C as friction with the atmosphere turned the kinetic energy of the craft into heat. Once again contact was lost with the ground as the intense heat turned the layer of atmosphere surrounding the module into a highly charged plasma through which radio waves could not penetrate. The command module emerged from its immolation travelling at about 480 km per hour at an altitude of 10.7 km. The final means of slowing the ship was to deploy three large parachutes from which the astronauts floated down into the sea to be collected by helicopter and flown to an aircraft carrier.

On the early lunar landing missions, the plan called for the astronauts to be transferred straight into an isolation chamber. Although scientists were quite confident that no micro-organisms could live on the Moon, it was deemed safest to keep the astronauts in quarantine for a period to see if any diseases developed. Consequently when they alighted from the helicopter onto the deck of an aircraft carrier they were dressed in full biological isolation suits and walked straight into a quarantine chamber where they lived for the next two weeks. This practice continued until Apollo 15. Later missions ended with the astronauts climbing down out of the helicopter in fatigues and baseball caps.

In all there were eleven manned Apollo missions, of which nine flew to the Moon. Apollo 8's mission was to fly the command and service modules into orbit round the Moon. Apollo 10 took a lunar module along as well and flew it to within nine miles of the surface. There were eight missions that flew with the intention of landing, although due to the accident on board Apollo 13 only seven landings took place. The original plan was to continue until Apollo 20, but declining public interest and American involvement with other budget-draining activities (Vietnam, for example) led to pressure on NASA and the programme ended with Apollo 17.

1.2 The Moon

As the brightest object visible in the night sky, it is hardly surprising that the Moon has been a major feature in mankind's imagination[3]. In 1865 Jules Verne wrote a story in which explorers travelled to the Moon in a shell shot from a giant cannon (*From the Earth to the Moon*). Although the technology is impossible, the theme of lunar exploration was established early in the history of science fiction.

The most conspicuous feature of the Moon's appearance from Earth is the manner in which it waxes and wanes in a regular monthly cycle. This is due to the way in which its position changes relative to the Earth and the Sun. One side of the Moon is always illuminated by sunlight and what we see from Earth is the changing perspective on that side as the Moon turns in its orbit about us.

It is easy to see that the surface features of the Moon do not change from day to day. All that happens is that we can see more or less of the same face as the Moon's crescent changes. This probably does not take most people by surprise—yet it is a curious fact. The Moon is an unusual shape—it is not round, rather it is slightly egg shaped with the pointed end directed towards the Earth. The action of the Earth's gravitational force on the egg-shaped Moon has, over the history of the system, made the time it takes for the Moon to turn on its own axis the same as the time taken for it to orbit the Earth. Without this we would see a changing face as the Moon turned on its axis. As it is, the similarity between the rates of these two 'rotations' ensures that the same face is always towards us.

The first human beings to see the far side of the Moon with the naked eye were the three astronauts who took part in the Apollo 8 mission—the first time that a spacecraft carrying humans had been placed in orbit about the Moon[4]. Interestingly, the three astronauts on Apollo 8 (Borman, Lovell and Anders) had names which were quite similar to those chosen by Verne for his three heroes on the first voyage to the Moon (Barbicane, Nicholl and Ardan). This was an irony that was not lost on the Apollo 8 crew. Neither was the fact that Verne left his characters stranded in orbit round the Moon at the end of his novel[5].

The two sides of the Moon are very different in appearance. The near side is dominated by large dark areas called *maria* (seas) and bright areas which are the lunar highlands. Without the aid of telescopes, ancient lunar observers thought that the dark areas were seas of water as on Earth. Consequently they named them with this in mind—for example *Oceanus Procellarum* (Ocean of Storms) or *Mare Serenitatis* (Sea of Serenity). Some of them are quite large. The Ocean of Storms is larger in area than the Mediterranean. In total the maria cover some 16% of the Moon's surface—mostly on the near side. We can easily tell that they represent the youngest areas of the Moon as they are relatively lightly cratered compared with the highly pockmarked highlands. Over the millions of years since it was formed the Moon has been bombarded with meteorites. Unlike the Earth, the Moon has no atmosphere to protect its surface from this battering. As a result the Moon's crust is fragmented and covered with craters which are the holes left by meteorites striking the surface. Sometimes when meteorites have struck, the material thrown up can still be seen as long streaks, or *rays*, spreading out from the crater.

About 4 billion years ago, several huge impacts occurred, forming vast craters. Sometime between 4 and 2.5 billion years ago active volcanoes caused lava flows which filled these impact basins with much darker coloured material forming the maria that we can now see. The lava not only filled the basins, it also covered up any old craters that had been formed in the area. Since then a few more meteorite impacts have marked the surface of the seas, but the highlands which escaped the lava floods show a complete history of bombardment dating back to the Moon's formation.

The rocks found on the Moon's highlands are at least 4.5 billion years old, which dates them as forming at about the same time as the Earth. This makes them the oldest rocks on the Moon. They are rich in a light-coloured mineral, called feldspar, which gives the lunar highlands their bright colour.

Not all meteorites are large enough to leave craters. Much smaller ones, called micrometeorites, have also continually rained down on the Moon over the millennia. Their bombardment has thoroughly pulverized the rocks on the

Figure 1.7 The near side of the Moon. The dark impact basins (maria) and the bright uplands are clearly visible, as well as some bright craters with ejecta rays.

surface, forming rock fragments and a fine dust-like debris which together are referred to as the *regolith*.

On Earth regolith is formed by the *weathering* of rocks. Weathering is a general term for all the processes that can break up rocks. These processes can be physical (such as water in rocks turning into ice which expands, causing them to crack), chemical (minerals dissolving in water or acids), or biological (such as plant roots widening cracks in rocks). Such processes cannot take place on the Moon due to the lack of atmosphere, water and wind.

The lunar regolith is found covering the whole of the Moon's surface, aside from steep crater and valley walls. It is 2 to 8 metres thick on the maria and possibly more than 15 metres on the highlands, depending on how long the rock underneath it has been exposed to bombardment.

The lunar far side is much closer in character to the highlands of the near side. Although there are basins comparable in size to some of the major maria on the near side, they are not filled with the same dark material. Lunar geologists have

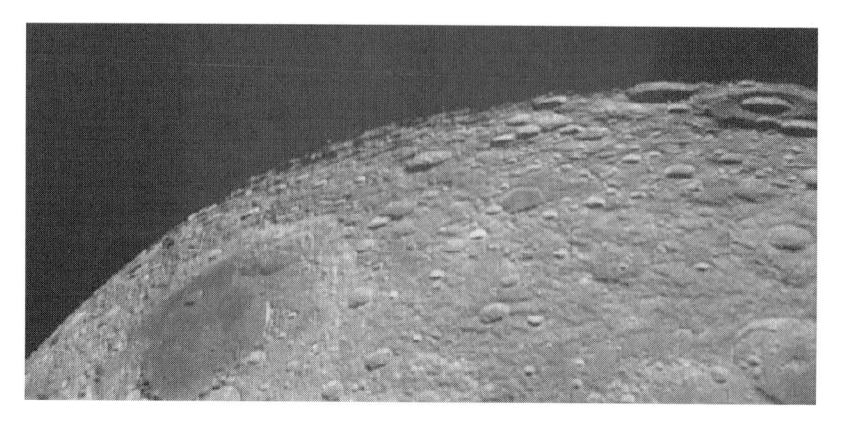

Figure 1.8 The far side of the Moon, as photographed by the crew of Apollo 13. The large, dark, smooth-looking feature on the left in this picture is known as the 'Mare Moscoviense'. First photographed by an early Soviet lunar probe, it is one of the few maria on the far side of the Moon.

discovered that the centre of gravity of the Moon is not quite at the Moon's geometrical centre—it is about 2 kilometres nearer to the Earth[6]. This may be because the Moon's crust is thicker on the far side. Crustal material is less dense than the underlying material (mantle): therefore, if there is a greater thickness of crust on the far side of the Moon's geometrical centre there is also less mantle than on the near side—hence the near side is heavier. This would tie in with the lack of maria on the far side, despite there being suitable basins to fill—with a thicker crust lava would have much greater difficulty making it to the surface.

Apollo's contribution to lunar science

Before the Apollo Moon landings, there was a great deal of speculation about the nature of the Moon, but little in the way of definite information. While they were on the surface, the Apollo astronauts carried out geological surveys and sampled rocks that were later brought back to Earth for analysis. As a result of their observations and sampling, Apollo was directly responsible for establishing a sequence of basic facts about the Moon.

- The Moon is made of rocky material. Various rocks show evidence that the lunar material has been melted, squirted out from volcanoes, and crushed by meteorite impacts. There is a thick *crust* (60 km), a fairly uniform *lithosphere* (60–1000 km), and a partly liquid *asthenosphere* (1000–1740 km) which correspond to equivalent layers found within the Earth. The Moon may have a small iron core. Some rocks show evidence that there was a magnetic field in the past, although no lunar field exists today.

- The Moon is extensively covered with craters which date from a variety of periods in the history of the solar system. The Apollo astronauts brought back rock samples that enabled various craters to be dated. This provided a key that can be used to establish timescales for geological evolution on other planets, such as Mercury, Venus and Mars, which show evidence for periods of crater formation similar to those found on the Moon. Now that we can date the craters on the Moon, we can make conclusions about the ages of similar craters on these planets. Before Apollo, however, the origin of lunar impact craters was not fully understood and the origin of similar craters on Earth was highly debated.
- Moon rocks range in age from about 3.2 billion years in the maria to nearly 4.6 billion years in the highlands. On Earth the various active geological processes (plate tectonics and erosion) continually recycle rocks from the oldest surfaces of the Earth. No such processes exist on the Moon, so it has far older surfaces still visible. The oldest rocks found on the Earth's surface are about the same age as the youngest rocks found on the Moon. Consequently the Moon has a far more extensive record of the events that must have affected both bodies.
- The Earth and the Moon have a common ancestry. Studying the amounts of various oxygen isotopes found in Moon rocks show that the proportions are very similar to those found in rocks on Earth. However, the Moon is highly depleted in iron and in the elements needed to form an atmosphere or water.
- Extensive testing revealed no evidence for life, past or present. In fact there was remarkably little evidence for standard organic chemicals that do not need to be formed biologically. Such compounds are regularly found in meteorites striking the Earth, and the little found on the Moon is probably due to such meteorites striking the Moon as well[7].
- There are three types of rock found on the Moon: *basalts*, *anorthosites* and *breccias*. Basalts are the dark lava rocks that fill mare basins. They are similar to the lavas from which the Earth's ocean crusts are made, but much older. Anorthosites are the light rocks from which the highlands are formed. They contain large amounts of the feldspar mineral. Breccias are composite rocks formed from all other rock types through crushing, mixing and being smashed by meteorite impacts. Significantly, there were no sandstones, limestones or similar rocks found on the Moon. Such rocks formed on Earth due to the action of flowing water laying down sediments.
- Early in its history (4.4–4.6 billion years ago), the Moon was molten to great depths. As this 'magma ocean' slowly cooled minerals started to crystallize within it. The earliest ones to form (those with a high melting point) were comparatively dense and so tended to sink as they formed. The low-melting-point material crystallized later near to the surface—being less

dense—and now forms the lunar highlands. Over billions of years meteorite impacts reduced much of the ancient crust to leave the mountain ranges that we now see between basins.

- The large, dark basins are gigantic impact craters, that were later filled by lava flows about 3.2–3.9 billion years ago.
- Large mass concentrations (mascons) lie beneath the surface of many lunar basins. These mark places where lava has formed a thick layer under the surface.
- The Moon's shape is slightly asymmetrical, probably due to the continual pull of the Earth's gravity affecting the way the Moon has evolved. Its crust is thicker on the far side, while most volcanic basins—and unusual mass concentrations—occur on the near side.
- The surface of the Moon is covered by a rubble pile of rock fragments and dust—the lunar regolith. The radiation pouring out from the Sun acts on the regolith and surface rocks of the Moon, changing their chemical composition. Consequently the Moon has recorded four billion years of the Sun's history. We are unlikely to find such a complete record on any other planet.

Origin of the Moon

There have been four basic theories put forward to explain where the Moon came from. Each of them has to deal with two significant facts concerning the Moon. Firstly, out of all the planet/moon systems in the solar system, our Moon is by some margin the largest compared to the planet it orbits. Secondly, the Moon is composed of a much smaller percentage of iron than is the Earth.

Each theory relies on the well accepted model of how the Sun and planets formed. About 4.5 billion years ago a large cloud of gas slowly contracted under gravity into the glowing ball that became the Sun. As the atoms in the cloud were pulled together by their mutual gravitational attraction they picked up speed (just as a falling rock will accelerate). Collisions with one another ensured that the atoms ended up moving about at random. In a gas the average kinetic energy of molecules in random motion is what we measure as temperature, so as the gas cloud collapsed the collisions ensured that its temperature increased. Eventually the temperature rose sufficiently to trigger the sort of nuclear reactions that power every star. At this point the energy produced by these reactions served to increase the temperature further and the cloud stopped contracting as the atoms were now moving fast enough to resist the inward pull of gravity (this is the natural tendency of any hot gas to expand). Currently the Sun is enjoying comfortable middle age, settled in the balance between the gravitational force of its mass trying to collapse it and the tendency of its hot gas to expand. This balance will continue for something like another 4.5 billion years.

As the central part of the gas cloud contracted to form the Sun, some heavier material left behind formed lumps of rock called planetoids. By colliding with one another and pulling on each other with gravitational forces, these planetoids eventually merged to form the rocky planets that now exist in the solar system. The surfaces of these planets were very hot and continually battered by meteorites, which prevented much cooling taking place. This is how the Earth formed.

One theory suggests that at this time the Earth was spinning very fast and threw off a large lump of material which formed the Moon (rather like mashed potato flying off a plate that is spun too quickly[8]). Provided this happened after the Earth's iron had sunk towards the centre of the forming planet (as it would do being heavier than the rest of the material), this could explain the lack of iron in the Moon. However, there is a considerable snag with this theory. If we study the motion of the Moon and calculate how fast the Earth's spin would be if it re-absorbed the Moon, then we find that the Earth would rotate on its axis once in 8 hours (rather than 24 hours). Unfortunately, although this is a very rapid rotation rate, it is not enough to account for throwing off a lump of material the size of the Moon. That would take a spin rate sufficient to shorten the day to something like 2 hours. Consequently, unless a great deal of energy has been lost in some fashion, the Moon could not have formed in this manner[9].

The second theory simply suggests that the Moon and the Earth formed at the same time out of planetoids, and rather than merging with the Earth, the Moon went into orbit round it. This offers no real explanation as to the very different metallic structures of the two bodies.

Another theory suggests that the Moon was formed somewhere else and was captured, in passing, by the Earth's gravity. This theory has two snags. The first is that the Moon would have to lose a great deal of energy in order to be captured by the Earth in this manner, and there is no obvious mechanism to account for this loss. The second point is that, aside from the iron content, the Earth and the Moon are rather too similar to suggest that they were formed in totally different places.

The fourth theory is the one that is currently given the most credence. This suggests that at an early point in the Earth's formation it collided with a very large object (probably about the size of the planet Mars). This massive impact vaporized the object and much of the Earth's crust. Some of the material thrown up by this fell back to Earth. Some of it escaped into space and the rest formed a ring of hot gas in orbit round the Earth. Eventually gravity pulled this material together to form the Moon. The nice aspect of this theory is that it comfortably explains the curious aspects of the Moon. If the collision took place after the Earth's iron-rich core was formed, then the outer material blasted away would be poor in iron—hence so would the Moon. Some of the material would be the remains of the colliding body, so the Moon would be similar in composition to the Earth, but not exactly the same.

Moon facts

Table 1.1

Mass (Earth = 1)	0.012
Radius at equator (Earth = 1)	0.27
Average distance from Earth (km)	384 400
Time to turn on axis (days)	27.321 66
Time to orbit Earth (days)	27.321 66
Gravitational acceleration at equator ($m\ s^{-2}$)	1.62
Average surface temperature (day)	107 °C
Average surface temperature (night)	−153 °C
Maximum surface temperature	123 °C
Minimum surface temperature	−233 °C

1.3 The immediate future

Currently the idea of space exploration is undergoing something of a popular revival. I do not think that there has been such excitement over matters relating to space since the early days of the Apollo programme. For various political and financial reasons space exploration entered something of a backwater after Apollo (with some exceptional successes that did not capture much public imagination). We only now seem to be moving out of that period. Interestingly, the rise of the Internet seems to have been a factor in this.

The Internet is a global connection of computers which the average home user with a PC (or Macintosh!) can tap into. It is a repository of information to which individuals at home, companies, research organizations and universities can all contribute. Much of the information is of no interest to an individual user (indeed, some of it seems to be of no interest to anyone!), but there are powerful computer programs called *search engines* that enable us to look for specific items amid the huge volume of information available[10].

In recent times, various teams working on satellites and space probes have made their data available on the Internet almost as quickly as it has arrived on Earth. If this data were solely in terms of dry numbers then it is doubtful that anyone

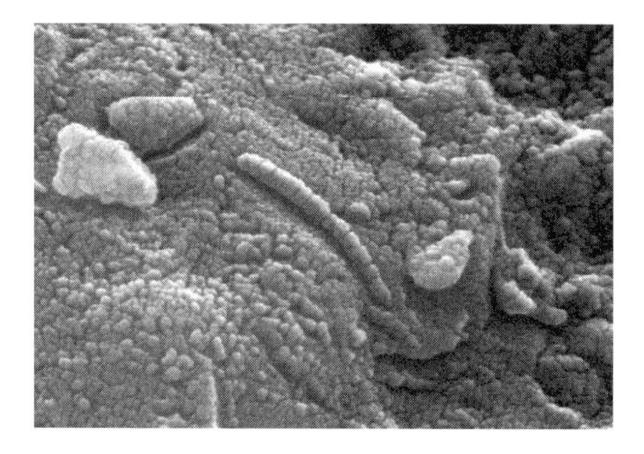

Figure 1.9 An electron microscope image of a sample of ALH84001. The worm-like structure in the middle may be a fossilized Martian life form.

but the professionals or highly skilled amateurs would be interested. However, some of the data has been in the form of staggeringly beautiful pictures. This has captured the public interest. The first photographs from the Galileo probe in orbit round Jupiter, the images from the Mars Pathfinder robot as it picked its way among large rocks on the surface (rocks given names like Yogi which seemed to become household names, even friends), the incredibly detailed images of the surface of Mars taken from orbit by the Mars Global Surveyor and the continual supply of breathtaking images from the Hubble Space Telescope have all caught the imagination. The computers on which these images have been stored for public access have regularly been flooded with 'hits' from home users all over the world. The Pathfinder website became the largest Internet event in history, with 566 million hits in the first month and a staggering 47 million hits on one day.

The public now feels far more involved with the process and in consequence more interested in what is going on.

Along with this, NASA announced in August 1996 the first tantalizing hints that life may have evolved somewhere other than on the surface of the Earth. The evidence for this claim came from the detailed study of a meteorite, unglamorously named ALH84001, found in the frozen wastes of the Antarctic. Incredibly, this meteorite seems to have originated on Mars.

About 15 million years ago a large meteorite impact on Mars blasted chunks of rock from the surface with such force that they escaped the gravitational pull of the planet. After further millions of years drifting in space, one chunk of rock crashed onto the Earth some 13 000 years ago, where it was covered in ice and preserved until found by an expedition in 1984. Detailed study produced some chemical evidence that bacteria had once lived within the meteorite. Fantastic electron microscope images of objects looking very suggestively like fossilized living forms again captured the imagination of the world.

Discoveries like this are so profound and so radical that they are not easily accepted by the scientific community. Indeed the team announcing the evidence spent two years carefully checking their results before daring to go public with the information. Many scientists are still sceptical and other explanations for the results have been put forward. However, one thing has become evident—the search for life on Mars has been pushed right to the front of the scientific agenda and with it the space programme has been given a much needed boost. It is now far more acceptable to start talking about a manned mission to Mars.

Now that the political motivation for space exploration has moved beyond competition with the 'red menace', perhaps the search for life (surely the most profound scientific endeavour we could possibly engage in) on the red planet will provide the key to loosening the purse strings of government.

Notes

[1] The original designation for the lunar excursion module had been LEM, but it changed to LM. Similarly common astronaut usage truncated the command and service module combination (CSM) to just command module (CM).

[2] Analysis that took place after the flight showed that the LM descent stage had 349 kg of propellant remaining of which about 45 kg would not have been useable. This would have given them about 50 seconds of hovering flight. The confusion over propellant was due to the propellant sloshing about in the tanks during the powered descent. This also caused a low propellant warning light to latch in the LM triggering a 94 second countdown to a 'bingo' alarm—which meant 'land in 20 seconds or abort'. Houston called up 30 seconds to bingo early due to the propellant slosh problems. The LM could still land in the 20 seconds after the bingo call provided that it was low enough to allow the engine to slow the descent to within the tolerance of the landing legs.

[3] One might have thought that the Moon was also the largest object in the night sky. However, one of this book's reviewers pointed out that some faint but naked-eye-visible galaxies cover a larger area of the sky than the Moon.

[4] The orbit was set to be 69 miles above the surface. The joke among the astronauts involved in running simulations for training purposes was 'wait until you see the 70 mile mountain on the far side!'

[5] Public outcry at this treatment was apparently so great that Verne had to write a sequel in which they made a safe return journey—*Around the Moon*.

[6] The centre of gravity of an object is its point of balance. If you balance an object on your finger, then the centre of gravity must lie on a vertical line through your finger. Imagine that I make a model of the Moon by taking two identically sized spheres and cutting them in half. One sphere is made from metal and the other from wood. If I join them together to make one sphere out of two different halves, then the geometrical centre would be the centre of the sphere, but the centre of gravity would be nearer to the heavy half.

[7] Organic compounds do not mean life. Organic chemistry is a branch of chemistry that deals with carbon compounds. Many carbon compounds are found in living systems, but they do not have to be. Rudimentary compounds out of which the complex molecules needed for life can be constructed have been found in gas clouds in space, in the atmosphere of Jupiter and in meteorites.

[8] You mean that you have not tried this?

[9] Other factors also come into account such as the angular momentum of the two bodies. Angular momentum is the momentum associated with rotational motion.

[10] Search engines are by no means infallible. When I was researching a previous book I wanted a copy of a diagram showing the distribution of galaxies in a slice of the universe. This had been produced by measuring the red shift of the light from the galaxies. I had seen this picture and knew that the blobs representing galaxies coincidentally drew the shape of a stick man. Consequently I searched for 'red, shift, galaxies, stick, man'. My first 'hit' was a car dealer in America wishing to sell me a red Galaxy with a stick shift...

Chapter 2

The best driver in physics

In this chapter we will consider some of the basic physics that we will need to apply in the rest of the book. The basics of force, mass and acceleration are developed. Newton's three laws of motion are discussed and applied to a variety of situations connected with space flight.

2.1 The first voyage to the Moon

The Apollo 8 mission was one of the most ambitious that NASA has ever undertaken. Originally planned as an Earth orbit test of the lunar module, delays completing that complicated craft left room for a change in plans.

The CIA had uncovered intelligence suggesting that the Russians were intending to send one of their manned *Soyuz* spacecraft to the Moon. Keen and ambitious engineers in NASA wanted to ensure that the American space programme was not left behind by the Russians once again. They lobbied to send the second manned Apollo to the Moon—and not just to loop round and return as the Russians intended, to go into orbit. Consequently, on 21 December 1968 Frank Borman, William Anders and James Lovell lifted off for the Moon on the first manned flight of the Saturn V booster. The mission was a spectacular success full of historical moments. People watching it live will never forget the astronauts reading from the book of Genesis in orbit round the Moon on Christmas day, or the first sight of the Earth rising over the Moon.

The equipment performed faultlessly and the flight home became almost boring for the three-man crew. At one point Collins (in mission control) mentioned to Anders that his son had asked who was doing the driving up there. Anders' dry reply was 'I think Isaac Newton is doing most of the driving right now'.

2.2 On spacecraft and shopping trolleys

You would think that the business of flying a spacecraft to the Moon was terribly difficult, requiring an astronaut to have his 'hands on the wheel' all the time. However, flying to the Moon is, in some ways, a good deal simpler than most people imagine. This is because the events that we see and experience around us govern our imaginations. In my experience things do not go in the direction that one wishes. I struggle with shopping trolleys.

Partly this is due to the fact that shopping trolleys lead such an abused life that the four wheels on most of them are all pointing in different directions. However, mostly the problem is friction. The constant presence of the force of friction distorts our thinking about how objects move. I suspect that most people, if asked, would think that you have to continually push an object in order to keep it moving. After all, that is what you have to do with shopping trolleys. Similarly, I imagine most people think that Apollo's engines had to keep on working otherwise the spacecraft would grind to a halt (many science fiction programmes and films get this wrong!). However, in space there is no friction. This makes a huge difference. The motion of objects is much simpler in space, but our intuition of how things ought to move is totally wrong. Newton was a man with a powerful imagination. He imagined what it would be like to move without friction and from this picture deduced the laws of motion that dictate how objects really move. Newton's laws are clearly visible in the 'simple' situation of being in space.

2.3 The power of imagination

Newton would have understood all the principles involved in navigating to the Moon and back. His laws of mechanics, which he first published in the 1600s, are perfectly adequate to calculate the orbits and trajectories required to navigate about our solar system, not just to the Moon and back.

In his book, *Principia Mathematica*, Newton described the principles that would guide the whole of physics right up to the start of the twentieth century. Newtonian mechanics was eventually replaced by the theory of relativity and by quantum mechanics between 1905 and 1930. However, relativity only differs from what Newton discovered when an object is moving at a speed close to that at which light travels—three hundred million metres per second[1]. No spacecraft has ever moved faster than a small fraction of this speed. Similarly, quantum theory only shows the limitations of Newton's mechanics when the objects involved are very small—typically the size of atoms, about 10^{-10} m across. Spacecraft do not fall into either of these categories.

The cornerstones of Newtonian mechanics are the three laws of motion first set out in their completely correct form in the *Principia* (Galileo had been close to formulating the same ideas, but lacked Newton's mathematical insight). The first two laws form a pair that tells us how things move with or without forces acting on them. The third law is about forces and how objects interact with each other.

Newton's laws of motion

- If an object does not have a net force applied to it, then it will remain stationary or if it was moving will continue to move in a straight line at a constant speed.
- If an object does have a net force applied to it, then the force will alter its motion either in direction or by changing its speed or both.
- If one object applies a force to another, then the second object will also apply a force to the first and this force will be the same size, but in the opposite direction to the first one.

These laws seem so simple it is a wonder that it took so long for someone to figure them out. The problem is that the first law in particular is contrary to what we seem to observe in everyday life. A shopping trolley pushed along and then released will roll across the floor, gradually slowing down until it comes to rest. Observations like this (but not with shopping trolleys in Newton's time) encouraged people to think, mistakenly, that a force was required to *keep* something moving. It took Newton's powerful imagination to extract the truth from the observations of daily life. He imagined an object moving through empty space and realized that if no forces at all were acting then there would be nothing to stop the object—it would just coast along at a constant speed. Equally, if an object had a set of forces acting on it which happened to cancel each other out, then it would be as if there was no force acting. For example, if two equal forces were pulling on a stationary object from opposite directions, then it would stay exactly where it was.

The phrasing of the laws of motion quoted earlier caters for this circumstance by the use of the term *net force*. The net force acting on an object is the force that results from adding up all the forces acting, taking both their sizes and direction of action into account.

If an object is just floating without moving at all, there would be no reason for it to start moving—something would have to push it to get it going (a net force), but once it was moving the force would no longer be needed. A shopping trolley in space could be pushed once, and then it would serenely coast along, moving at a constant speed in the direction it was pushed.

We will see in chapter 4 that Newton's powerful imagination also enabled him to see how an object could stay in space permanently orbiting the Earth.

The second law is also deceptively simple. In school physics it is often written in the form of a mathematical equation:

$$\text{force} = \text{mass} \times \text{acceleration}$$

acceleration being the rate at which velocity is changing[2]:

$$\text{acceleration} = \frac{\text{change in velocity}}{\text{time elapsed}}.$$

For example, my favourite car, the Aston Martin DB7, can go from rest to 62 miles per hour in a shade under 6 seconds. This is an acceleration of:

$$\text{acceleration} = \frac{62 \text{ mph} - 0 \text{ mph}}{6 \text{ s}} = 10.3 \text{ mph every second.}$$

In other words, the speed of the car increases by 10.3 miles per hour every second while it is accelerating. Now, these are very clumsy units as far as a physicist is concerned. They prefer to work in metres so that speed is measured in metres per second. Using these units the calculation is:

$$\begin{aligned} 62 \text{ mph} = 100 \text{ km per hour} &= 100\,000 \text{ m per hour} \\ &= 100\,000 \text{ m per 3600 seconds} \end{aligned}$$

so

$$62 \text{ mph} = \frac{100\,000 \text{ m}}{3600 \text{ seconds}} = 27.8 \text{ metres per second}$$

$$\begin{aligned} \therefore \text{acceleration} &= \frac{27.8 \text{ metres per second}}{6 \text{ s}} \\ &= 4.63 \text{ metres per second every second.} \end{aligned}$$

So, every second the car increases its speed by 4.63 metres per second. Physicists like to write this as 4.63 m/s^2 (or 4.63 m s^{-2}). Note that this does not mean that we are dividing by a 'square second' (whatever that is!), although it is pronounced 'metres per second squared'—which adds to the confusion.

In contrast to the Aston Martin, the Saturn V rocket accelerated off its launch pad at 1.92 m s^{-2}. Now, this seems to be a very small acceleration. All the images we see of spacecraft lifting off suggest that there are vast forces at work. Surely these forces cannot result in the ship crawling off the launch pad? The problem here is that we are underselling the size of the forces needed to achieve even this crawl[3].

Newton's second law allows us to calculate how great a force is required to achieve a given acceleration:

$$\text{force required} = \text{mass of car} \times \text{acceleration required.}$$

In this formula, if we use mass in kilograms (kg) and acceleration in metres per second squared (m s^{-2}) then the force automatically comes out in newtons (N)—the unit of force named after Sir Isaac Newton. To get a feel for the size of a 1 N force, remember that a 1 kg mass has a weight[4] of 10 N.

We can now calculate the force that the engines of the Saturn V booster needed to develop in order to lift the Apollo spacecraft off the ground. As it lifts off, a rocket is not moving fast enough for air resistance to be a problem (although it will be later in the flight to orbit), but the engines have to overcome another force which is considerable. Saturn Vs lifted off vertically, so the full weight of the rocket was acting downwards, opposite to the direction in which it was trying to accelerate. In order to move an inch off the ground, the engines needed to develop a force at least as big as the weight of the rocket[5]—28 500 000 N.

Fully laden with propellant the Saturn V had a mass of 2 900 000 kg (two thousand nine hundred tonnes[6]), so to accelerate off the launch pad at 1.92 m s^{-2} the force required was:

$$\begin{aligned}(\text{force of engines} - \text{weight of rocket}) &= (2\,900\,000 \text{ kg}) \times (1.92 \text{ m s}^{-2}) \\ &= 5\,570\,000 \text{ N}\end{aligned}$$

making the force of the engines:

$$28\,500\,000 \text{ N} + 5\,570\,000 \text{ N} = 34\,100\,000 \text{ N} \qquad \text{(to 3 significant figures)}^7.$$

During launch this mighty rocket developed 85 times as much power as the Hoover Dam.

2.4 Falling

At the end of one of the Apollo 15 Moon walks, astronaut David R Scott carried out a short demonstration in front of the TV cameras. The demonstration is familiar to teachers of physics, but Scott had the advantage of doing it on the Moon. On the Moon, the gravity is so low that no atmosphere remains. No atmosphere means no air resistance. No air resistance means that it is easier to demonstrate one of the most important features of how gravity works.

The demonstration Scott carried out was to drop a falcon feather (the lunar module for that mission had been named *Falcon*) and a geological hammer from the same height. Despite being obviously different in weight they struck the lunar surface together. The pull of gravity on the Moon is less than that on Earth, but the physics is the same—all objects on the Moon accelerate under the action of gravity at the same rate. On Earth the much greater air resistance masks this effect.

When an object is dropped it accelerates towards the ground as the Earth is pulling on it with the force of gravity—the force we call the weight of the object. The size of this force is determined by the mass of the object—the greater the mass the bigger the weight.

Now, Newton's second law of motion tells us that the bigger the mass of an object the smaller the acceleration that will be produced by a given force. This is an extraordinary thing. The law of gravity says that the bigger the mass of an object, the greater is its weight. When an object is falling, neglecting air resistance, the only force acting is its own weight (technically, this is called *free fall*). The resulting acceleration does not depend on its mass at all.

Compare two falling objects. The one with a small mass will have a small force acting (its weight), but a small force can still produce quite an acceleration on an object with little mass. The heavier falling object has a bigger force acting on it, but as it is a larger mass, this bigger force will not produce any more acceleration on the object than that experienced by the small mass falling next to it.

Another way of showing that in free fall all objects accelerate at the same rate is to use the following mathematical argument:

$$\text{weight of object} = \text{mass} \times \text{strength of gravity field}$$

$$\text{acceleration of object} = \frac{\text{force on object}}{\text{mass of object}} = \frac{\text{mass} \times \text{strength of gravity field}}{\text{mass}}$$

$$\therefore \text{acceleration of object} = \text{strength of gravity field.}$$

At the equator, the strength of the Earth's gravity will accelerate an object at the rate of 9.8 m s^{-2} (we have been rounding this up to 10 m s^{-2} so far). Another way of saying this is that the strength of the Earth's gravity at the equator is 9.8 m s^{-2}. This acceleration, known as g, is a convenient way of measuring large accelerations (such as that experienced by astronauts during lift off). For example an acceleration of $10g$ would be 98 m s^{-2}.

It is easy to see that the force required to produce an upward acceleration of $10g$ would be 11 times the weight of the object. The force has to overcome the weight of the object and then produce an upward acceleration of $10g$, so:

$$(\text{force} - \text{weight of object}) = \text{mass} \times \text{acceleration} = m \times 10g$$

$$\therefore \text{force} = m \times 10g + \text{weight of object}$$
$$= 10mg + mg = 11mg = 11 \times \text{weight of object}.$$

Astronauts experiencing $4g$ during launch would feel five times heavier due to the acceleration[8]. Fighter pilots can 'pull' up to $8g$ when they are banking their aircraft into a tight turn or roll. Formula 1 racing cars often have accelerations of $3g$ when they are turning corners.

The practical upshot of this property of gravity, that all objects fall at the same rate, is that an astronaut in a spacecraft orbiting the Earth feels 'weightless'[9] as he and the craft he is floating in are both falling towards the Earth at exactly the same rate. How orbits allow objects to fall without hitting the ground is the subject of chapter 4.

2.5 Forces during lift off—the astronauts

Anyone who has seen film of astronauts during lift off will recall them being 'pressed down' into their acceleration couches, or perhaps even the skin on their faces being distorted and 'pulled down' over the bones of their cheeks. Much of this is the artistic licence used in science fiction films to make the scene more dramatic, but real film shows that the basic effects are present when a large rocket booster takes off.

This is due to the forces acting on astronauts during lift off. However, even though they describe it as like being 'pressed into the couch by a giant hand' a careful consideration of what is happening shows that the effect is actually due to an upward force pushing them off the ground.

Figure 2.1 shows a stylized astronaut lying on an acceleration couch. The forces acting have been drawn in as arrows.

Notice that there are only two forces acting directly on the astronaut—his weight and the force that the couch applies to him. The engines of the rocket do not apply a force directly to his body as he is not connected directly to them. He is strapped into an acceleration couch. The couch is (presumably) bolted down and so experiences a force from the engines transmitted via the body of the spacecraft. Due to the action of this force the couch will start to accelerate upwards. The astronaut, however, does not immediately accelerate with it. At the moment that the rocket starts to move, the size of the force acting on the astronaut from the couch is just equal to his weight—after all the couch is preventing him from falling to the ground. As the rocket starts to move the couch accelerates past him. The material of the couch is distorted and he appears to

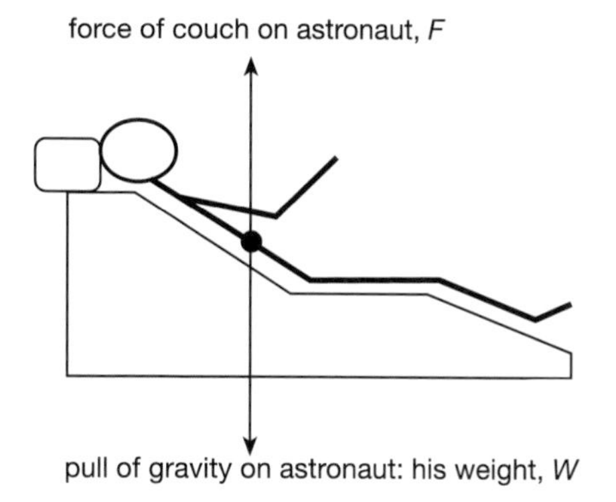

Figure 2.1 Forces acting on an astronaut during lift off.

sink into it, not because he is being pushed down, but rather because the couch is trying to overtake him! As the material distorts more, so the force that it applies to the astronaut will increase until it is sufficient to make him accelerate upwards at the same rate as the couch. At this point he will stop sinking further into the couch and the distortion of the materials remains constant.

The same effect causes the faces to be distorted. The skin of the face is not rigidly attached to the bone beneath, so the head tries to accelerate through the skin, causing the distortion.

If this still does not seem very plausible, then imagine that instead of lying on a couch the astronaut is standing on a weighing machine (figure 2.2). Weighing machines work by measuring the amount by which a spring is compressed when someone stands on it. The heavier the person, the greater the compression of the spring. As the spring compresses, the force that it applies upward to support the person increases. The compression stops when the upward force is equal to the person's weight.

As the rocket starts to accelerate, the force acting up on the spring increases but the force acting down remains the same as this is due to the weight of the astronaut. In fact, this force never changes. The two forces acting on the spring are not equal, which is why the spring starts to accelerate upwards. However, at first the astronaut does not. The two forces on him are still equal. This means that the bottom of the spring is moving upwards while the top is not, which is what causes the spring to compress. As the spring compresses so the force which it exerts upwards on the astronaut increases. The forces on him are no longer equal and he also starts to accelerate upwards. The spring continues to compress until the force acting on the astronaut is enough to make him accelerate upwards at the same rate as the rest of the rocket. The end

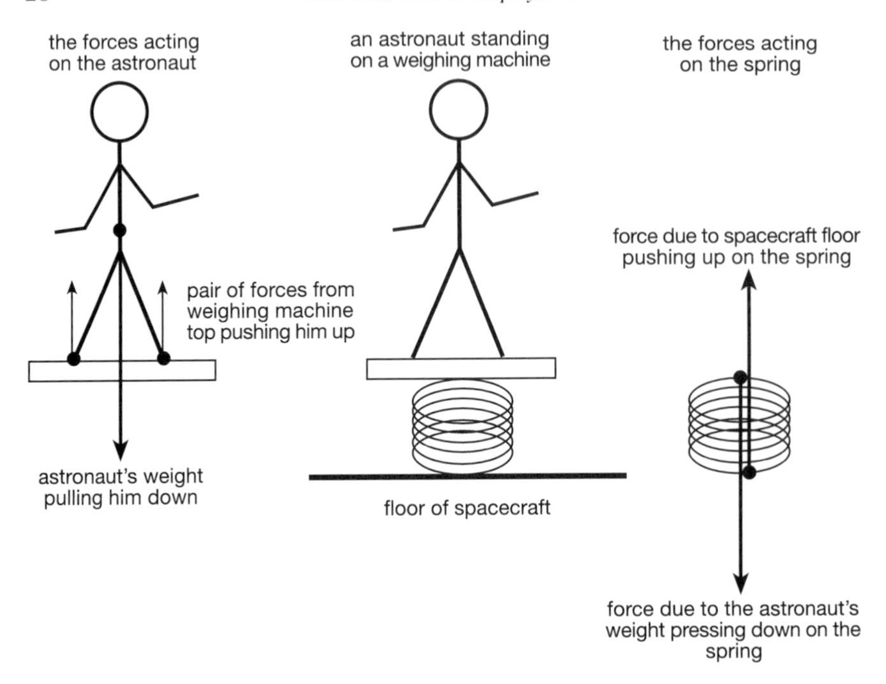

Figure 2.2 Forces acting on a weighing machine.

result is that the weighing machine records a greater value (as the spring is now compressed more) yet the force of gravity has not changed. We will return to this when we discuss weightlessness in chapter 4.

The couch behaves just like the spring (it may after all have springs inside it). In fact, all objects behave a little like the spring. When you walk on the floor it distorts a tiny amount beneath you to support your weight. A plank resting across two oil drums will bend in the middle to support someone standing on it. It is all the same idea.

During lift off the astronauts report feeling heavy and say that it is much harder to lift their arms to operate the switches on the control panel above them. During lift off they feel heavier, in orbit they feel weightless, yet it is hard to believe that the force of gravity on them has changed[10].

Consider raising an arm resting on the acceleration couch to press a switch on the control panel in front of the astronaut's face. With the arm resting on the couch the force needed to accelerate the arm is coming from the material of the couch. Once the astronaut has lifted his arm off the couch, the accelerating force must come from his muscles. He must achieve an acceleration slightly

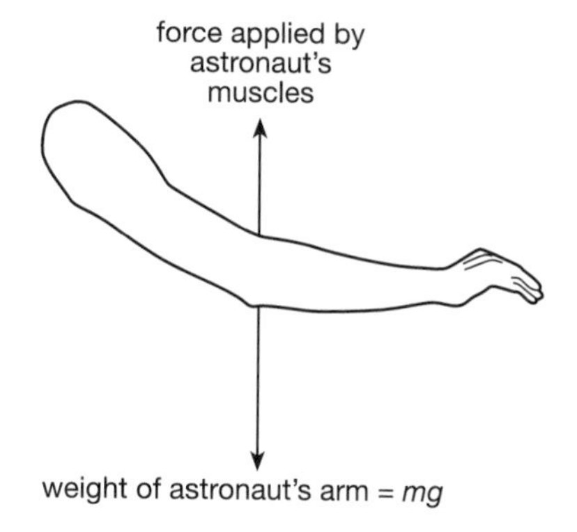

Figure 2.3 Forces acting on an astronaut's arm.

greater than that of the spacecraft otherwise the couch will catch up with his arm (from the astronaut's point of view, his arm will fall back). In order to achieve this acceleration he must apply a force much greater than the weight of his arm. This is shown in figure 2.3.

With the spacecraft accelerating upward at $4g$, the total upward force on the arm must be enough to provide this acceleration:

$$\text{mass of arm} \times 4g = (\text{upward force from muscles} - \text{weight of arm})$$

$$\therefore \quad 4mg = F - mg$$

$$\therefore \quad F = 5mg$$

so the astronaut has to apply a force five times greater than the weight of the arm—which is why it seems five times heavier.

2.6 Forces during lift off—the spacecraft

The image of a rocket lifting off the launch pad is a powerful one. Ask most people what is happening and you tend to get an answer along the lines of the rocket 'pushing' against the ground.

Figure 2.4 shows one of the Apollo spacecraft (Apollo 15) being launched by a Saturn V booster. The column of burning exhaust gas strikes the ground and seems to spread out in all directions. Actually a significant portion of the clouds seen to either side of the launch tower is the steam produced by 106 000 litres of water per minute being used to cool the flame buckets absorbing the exhaust under the launch tower.

Figure 2.4 The launch of Apollo 15.

This image conjures up comparisons with objects pushing against the ground to lift upwards (like a person getting out of a chair). However, if you consider the next photograph (figure 2.5), showing a Saturn V in flight, then it becomes difficult to see how engines can work in space where there is nothing to push against.

Interestingly, this way of thinking is so ingrained that even the pioneer of practical rocketry Robert Goddard had to contest it. Goddard is one of the most important figures in the development of rocketry. In a series of experiments during the 1920s he established the principles on which boosters such as the Saturn V were developed. However, much of his work was not appreciated at the time. In a famous editorial in the *New York Times*, Goddard's work was criticized on the following lines:

> 'Professor Goddard does not know the relation of action and reaction, and the need to have something better than a vacuum (of space) with which to react... Of course he only seems to lack the knowledge ladled out daily in high schools.'

The editor's comment about action and reaction refers to the third of Newton's laws of motion. As a simple example of this law at work, consider what happens when you get up out of a chair.

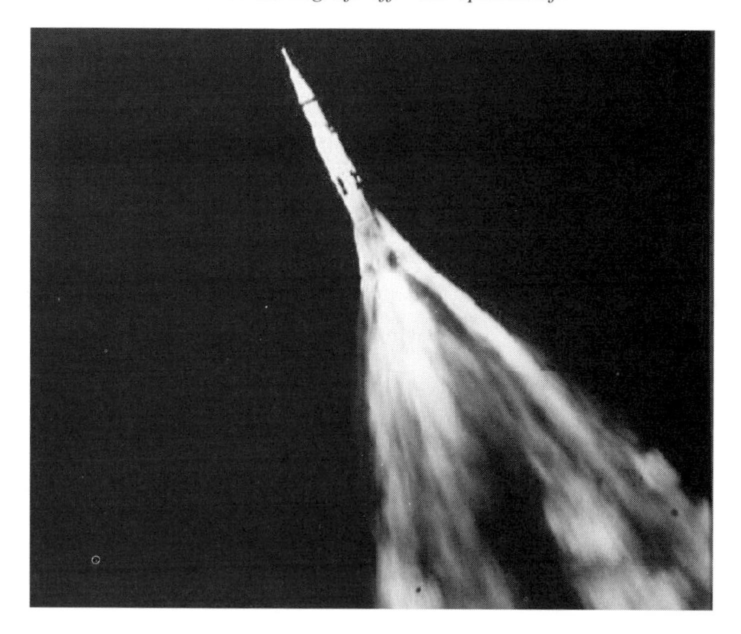

Figure 2.5 Apollo 6 in flight. The flame column now stretches some 800 feet behind the rocket.

When rising from a chair, you apply a force to the chair (the action force). Imagine putting your hands on the arms and pushing *down*—yet you go *upward*. It is *not* the force that you apply to the chair that moves you. Newton's third law of motion suggests that the chair, in response to the force that you apply, pushes back on you (the reaction force). It is *this* force, the force of the chair on you, that makes you move.

Similarly, when you set off across the room it is not a force that you apply to yourself that sets you moving. Consider carefully what is happening. As you start the first pace your leading foot lifts, moves forwards and makes contact with the floor. At this point you push backwards with that foot. The floor pushes back on your foot with a force that is forwards and it is this force that propels you. Imagine hanging from the ceiling by a string. All the pushing and waving of feet and arms that you can imagine will not result in you moving backwards or forwards. This is because you are not in contact with anything. Your limbs cannot apply a force to an object (like the floor or the chair) in order to trigger a reaction force on you by Newton's third law.

Astronauts carrying out space walks face the same problem. Freely floating in space, there is nothing on which they can push in order to trigger a force pushing on them.

This is the source of the confusion about rockets. Pictures showing take off seem to show the rocket pushing on the ground, and so we imagine that the rocket's motion is due to the ground pushing back (if we have thought about it at all). This is why people such as the editor of the *New York Times* thought that rockets could not work in space—there would be nothing to trigger a reaction.

Yet rockets do not work in this way at all.

Newton's third law applied correctly to rockets concerns the exhaust and the rocket's body, not the exhaust and the ground. The relationship is not straightforward and is more easily studied once you have some understanding of *momentum*.

2.7 Momentum

Increasingly these days we see pictures being beamed down to Earth of astronauts working in space. Perhaps the most famous incident in recent times was the repair of the Hubble Space Telescope carried out over several days by a team of astronauts working from one of the space shuttle orbiters. Successes like this tend to make the whole process of working with tools and objects in space seem very easy. In reality it is very far from easy and many of the tools have to be specially designed to work in space.

Consider a simple screwdriver. On Earth such a tool would be used to undo a screw. The process is very simple. You apply a force to the screwdriver, which in turn applies it to the screw loosening it in the process. The screwdriver works like a lever so that the force it applies to the screw is greater than the force that you apply to the screwdriver. However, this apparent simplicity hides the vital role that friction plays in the process. In space, where there is no air resistance and the astronauts may not be standing on anything, Newton's third law has some awkward consequences. If an astronaut applies a force to the screwdriver, and hence to the screw, then a force will be applied to him in return. To make the screw revolve one way, the astronaut is forced to revolve in the opposite direction!

On Earth the person using the screwdriver is in contact with the ground and the force of friction transfers the tendency to rotate through to the Earth itself.

Tools for use in space have to be specially designed to prevent opposite motion being passed on to the astronauts. However, this is not the only hazard involved in working in space. The apparent weightlessness of objects can lead to mistakes being made. On Earth a large box of electronics is both heavy (has a large mass so that the Earth is pulling on it with a large force of gravity—it has a great weight) and awkward to move (it has a large mass so that a large force is required to make it accelerate). In space, the box does not appear to have any weight—it floats in front of you. Consequently you get quite a surprise when you try to push it along. In space it still has a large mass and so requires a large force to get it moving. Unfortunately it also requires a large force to stop it as well!

In both cases, applying the necessary large forces to the box will result in the box applying equally large forces back (Newton's third law again). The problem is that if the astronaut is less massive than the box, then these forces will have a greater effect on the astronaut than the astronaut did on the box!

Imagine that astronaut A pushes a massive box of electronics towards astronaut B. As a result of A applying a force to the box, the box pushes A back and so A drifts away from the box. As the box arrives at B, she puts out her hands to slow the box down. The force that she applies to the box does slow it down, but unfortunately the force that the box applies to her speeds her up! The end result is that both start coasting along in the same direction in which the box was originally moving.

Sometimes the laws of physics hurt.

The situation of the two astronauts and the box can be fully analysed by considering the *momentum* of each object. Momentum is a useful quantity that often allows physicists to see easily what is happening in a situation when the details are very complicated.

The momentum of an object is defined as its mass multiplied by its velocity:

$$\text{momentum} = \text{mass} \times \text{velocity}$$
$$P = m \times v$$

which means that any object that is not moving has zero momentum.

Newton's laws can be used to show that the total amount of momentum in the whole universe never changes. If we were able to add up the momenta of every single object in the whole universe now and also able to do the same impossible sum one million years from now (or a million years ago for that matter), then we would get exactly the same answer.

One of the remarkable things about Newton's laws is that they allow us to make statements like this without even being able to add up all the momenta in the universe. That being the case, how can we be sure that the total never changes? The answer is to investigate what happens whenever objects collide. Our example of the two astronauts and the box is exactly the right sort of situation.

When we start off, none of the objects are moving. So the total momentum of this set of objects is zero.

Let us say that astronaut A pushes on the box with a force of 50 N for 0.1 s. The box has a mass of 200 kg, so it will accelerate at the rate of:

$$\text{acceleration} = \frac{\text{force (N)}}{\text{mass (kg)}} = \frac{50 \text{ N}}{200 \text{ kg}} = 0.25 \text{ m s}^{-2}.$$

An acceleration of 0.25 m s^{-2} for a time of 0.1 s results in an increase in speed of:

$$\text{acceleration} = \frac{\text{change in speed}}{\text{time}}$$

$$\therefore \text{change in speed} = \text{acceleration} \times \text{time} = 0.25 \text{ m s}^{-2} \times 0.1 \text{ s} = 0.025 \text{ m s}^{-1}.$$

As this is the change in speed, and the box was originally not moving, we conclude that the box ends up coasting away from astronaut A at a gentle speed of 0.025 m s^{-1}.

Consider now what happens to astronaut A. He is also acted upon by a force of 50 N, but the difference is that he and his space suit have a total mass of 100 kg. As a result, he accelerates at a rate of 0.5 m s^{-2} for 0.1 s and so ends up moving away from the box with a speed of 0.05 m s^{-1}. What of the total momentum?

Momentum of box = mass $\times$ speed = 200 kg $\times$ 0.025 m s^{-1} = 5 kg m s^{-1}.

Momentum of A = mass $\times$ speed = 100 kg $\times$ 0.05 m s^{-1} = 5 kg m s^{-1}.

At first this does not look like it is going to add up to zero, which is what the total momentum was originally. Then we realize that the box and astronaut A end up *moving in opposite directions, so perhaps we should count one of them as negative*. It does not matter which one, so we will take the motion of the box as positive:

$$\begin{aligned} \text{total momentum} &= \text{momentum of box} + \text{momentum of A} \\ &= (5 \text{ kg m s}^{-1}) + (-5 \text{ kg m s}^{-1}) = 0. \end{aligned}$$

Now, the collision between the box and B is not so easy to analyse. We cannot be sure about the size of the force acting. This will depend on a lot of factors, including how B moves her arms as the box hits her (people often move their hands and arms back as they catch a ball—this minimizes the forces acting—the same principle is at work when car designers make the front of cars crumple on impact). However, we can use momentum principles to see what the result will be.

If the result of the collision between B and the box is that they both end up coasting along together, then we can be sure that their combined momentum will be 5 kg m s^{-1}. If not, the total momentum in the universe would have changed when they hit each other. This is the power of momentum analysis. It enables us to study the end result of a collision or interaction between objects without necessarily understanding the details. In this case, once we are confident that the total momentum does not change, then we can say for certain what the final momentum of the box and astronauts must be.

There are many situations in which momentum does not seem to be a conserved quantity. For example, consider a simple situation such as a person suddenly deciding to start walking. That person's momentum has increased without any obvious decrease for another object. In fact there is a corresponding decrease. In order to start moving the person has applied a force to the ground, this alters the momentum of the Earth. In response the Earth pushes back, altering the person's momentum. There is a transfer of momentum from one to the other. In this case, as the Earth is so much more massive than a person, the speed change of the Earth is so tiny it can never be measured.

Now that we have some understanding of how momentum can be used, we can apply it to rocket motors.

2.8 The physics of rocket motors

We can gain some insight into how rocket motors provide propulsion by considering the following situation. An astronaut finds herself adrift having accidentally let go of the spacecraft during a space walk. In reality this is not a realistic situation as commonsense safety precautions involve the astronauts being tethered to the spacecraft[11]. After some experimentation she finds that she cannot make any progress by waving her arms or kicking her legs or making swimming motions. She cannot apply a force to anything, so there is nothing to apply a force to her to trigger her movement.

Fortunately she has been well educated in the laws of physics, so after some moments' thought removes a large spanner from her tool belt and throws it away as hard as she can (but not at the spacecraft!). As a result she starts to drift through space in the opposite direction.

Why has this happened? Considering the momentum we can see that the initial momentum of the spanner plus astronaut was zero (neither was moving). After being thrown the spanner gained momentum, and so to conserve the initial zero momentum the astronaut must also gain momentum, in the opposite direction.

In this case the forces involved are also quite simple. The astronaut applies a force to the spanner; by Newton's third law the spanner applies a force back. The force of the spanner on the astronaut sets her moving.

The astronaut stranded in space only has one spanner. Although she could remove other items and throw them away (up to a point) it would be far better to provide her with some continual source of mass that she can use to produce propulsion. Let us give her a large balloon that has been inflated to a high pressure[12]. (If this were a science fiction film, we could heighten the drama by having her use the hose that supplied oxygen to her spacesuit.) By carefully opening the end of the balloon she can release some of the gas held inside and as a result propel herself across space. With a certain amount of ingenuity she can even use the balloon to steer herself by pointing it in various directions. Although our intuition tells us that this will work, we need to stop and think about the physics of the situation.

From a momentum point of view all seems quite simple. The initial momentum was zero. The escaping gas gains a momentum in one direction and so the astronaut gains an equal momentum in the opposite direction. But this is not quite right. How did the escaping gas gain momentum? Opening a hole in the side of the balloon does not apply a force to the gas to give it momentum. It makes more sense to think that the gas that is escaping already had its momentum inside the balloon.

The balloon initially has zero momentum because all the molecules inside it have momentum and at any moment there is as much momentum in one direction as there is in the opposite direction. The gas molecules are moving about in a random way inside the balloon, which ensures that any unevenness in their motion cancels out very quickly. The internal momenta always add up to zero. Consider a simple case with only two molecules. One is moving towards the left, the other towards the right. When the molecules hit the far ends of the balloon they bounce off the skin and head back in the opposite direction. This continues to happen back and forth, always keeping the momentum total zero.

The forces acting in this situation are quite interesting. When a molecule strikes the skin of the balloon it applies a force to the skin. By Newton's third law the skin applies an equal force back and the molecule reverses direction. Equally, the skin is pushed out in the direction that the molecule was first moving. At the other end of the balloon the same thing is happening. The skin at this end is

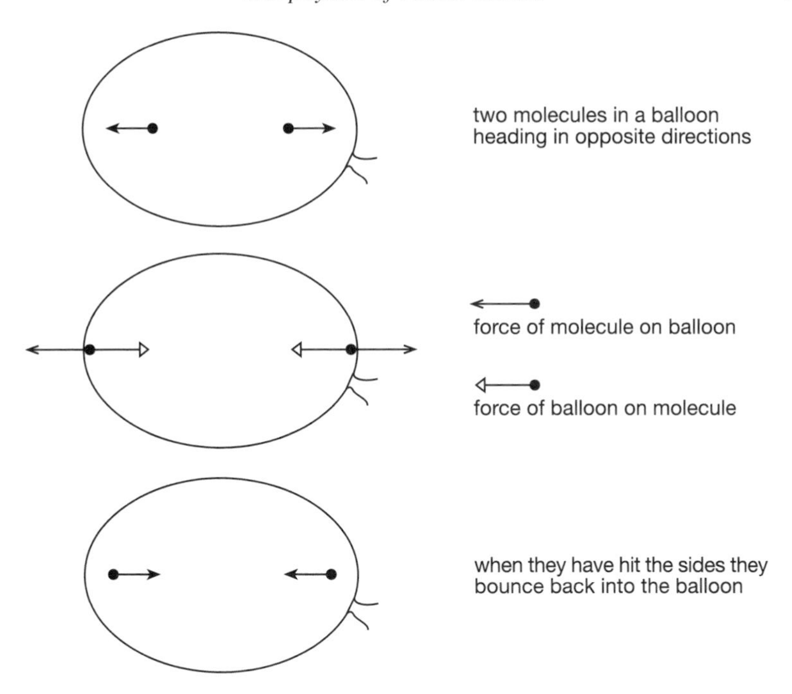

Figure 2.6 Two molecules bouncing back and forth in a balloon.

also pushed. This is why the balloon does not move—it is being pushed equally in both directions by the molecules striking the skin.

Now let us remove one of the sides of the balloon (figure 2.7). The molecule heading this way does not bounce off; instead it flies out of the balloon. It does not apply a force at one end of the balloon so now there is nothing to balance the force produced by the molecules at the other end of the balloon. It is this unbalanced force that causes the balloon to move forwards as gas escapes from the other end. This is the secret of rocket propulsion.

The final piece that we have to add is the generation of pressure without having to inflate a balloon to start with. We could imagine a cartoon rocket being powered by a large balloon inflated on the launch pad, but the engineering of such a situation would be problematic at best.

The answer is to use a chemical reaction that can generate high-speed molecules. In a rocket two chemicals are mixed which react with each other to provide a great deal of heat. This is very much like a controlled, continuous explosion taking place. The first stage of the Saturn V used kerosene (lighter fluid) and liquid oxygen. In the case of a lighter, a spark provides a localized region

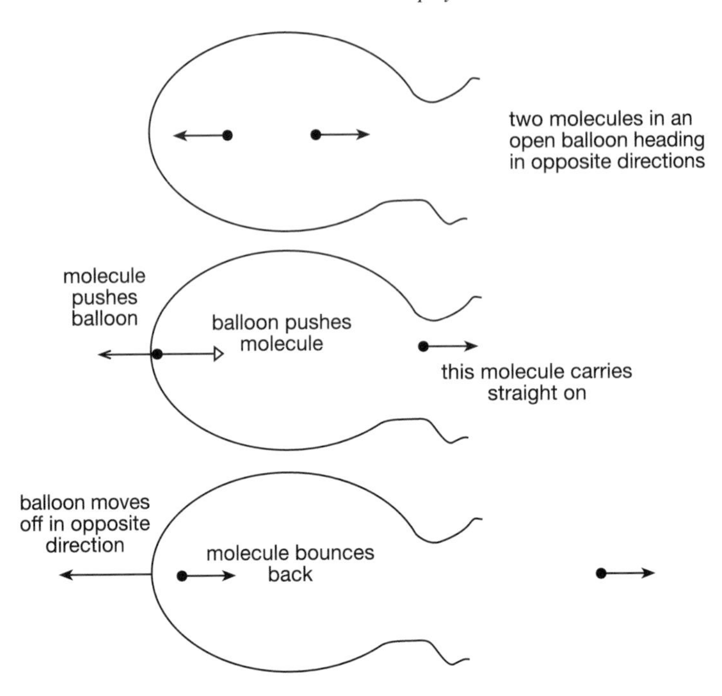

Figure 2.7 Molecules in an open balloon.

of high temperature within the liquid fuel. This triggers a chemical reaction between the fuel and the oxygen in the air. This reaction produces a great deal of heat which spreads in the fuel and a sustained burning is produced. The Saturn V first-stage motors used liquid oxygen rather than air, and used a secondary chemical that spontaneously ignited on contact with oxygen in the air to trigger combustion of the fuel, but the principle is the same.

The products of the reaction have a great deal of energy—their molecules are moving at very high speeds. The fuel and the liquid oxygen are mixed in a reaction chamber (balloon) in which the pressure builds up. At one end of the reaction chamber is a hole which allows the products to escape. The pressure on the reaction chamber is unbalanced and propulsion is produced. As long as there are chemicals left to react the rocket will continue to provide propulsion.

The figures for the Saturn V first-stage engines are staggering. In flight the five engines powering the stage between them consumed more than 4900 litres of kerosene and over 7500 litres of liquid oxygen per second, generating temperatures in excess of 3200 °C. In total they produced a force of more than 34 000 000 N.

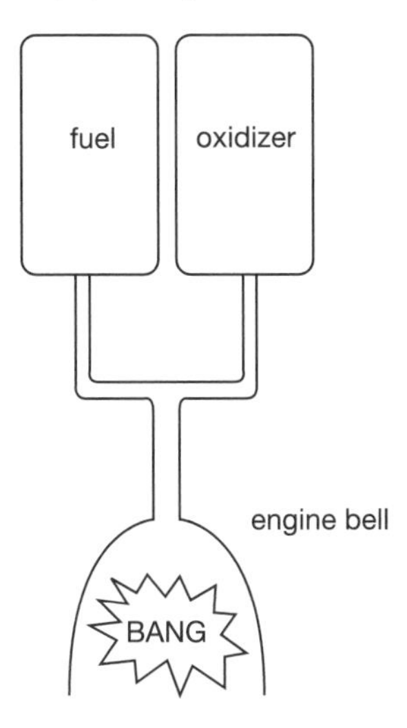

Figure 2.8 Principle of a chemical rocket.

Notes

[1] This comment specifically refers to Einstein's special theory of relativity. His general theory, published 11 years later, extends the special theory to include gravity. This was a further modification to Newton's ideas, but it only becomes really important when we are dealing with objects very much more massive than the Sun.

[2] Physicists like to distinguish between speed and velocity. Velocity is the rate at which something is moving in a specific direction. So, an object moving in a circle could be travelling at a constant speed, but its velocity would be changing as its direction of motion altered as it swung round.

[3] Also, as the rocket gets lighter (as propellant is consumed) the acceleration increases. By the time that the first stage is jettisoned the Saturn V is accelerating at something like 25 km s^{-2}.

[4] Weight and mass are often confused. The mass of an object is difficult to define at an elementary level, but is basically a measure of the amount of matter in an object. Weight, on the other hand, is a force—it is the force with which the Earth pulls on an object. This depends on how much matter the object contains and also the strength of the Earth's gravity. At the equator the Earth pulls on every kilogram mass with a force of 9.8 N—i.e. approximately 10 N.

[5] There is no such thing as a typical Saturn V. The figure used here is that quoted for the mass of the rocket at first motion in the Apollo 11 press kit. This mass included the ice formed on the metal surface due to the cold of the liquids stored, the nitrogen used to pressurize the propellant tanks, etc. For this reason it is not the same as the sum

of the various figures for stage masses and Apollo spacecraft masses quoted elsewhere in this book. It is quite difficult to be consistent about masses at various stages of the countdown and launch.

[6] This is the metric tonne (1000 kg), which corresponds to 1.1 US tons.

[7] The figures on which this calculation are based include the thrust at launch quoted in the Apollo 11 press kit i.e. 34 044 342 N.

[8] The peak acceleration experienced during a Saturn V launch was just before the first stage was jettisoned and was about $4g$.

[9] Weightlessness can be a confusing term—for starters it does not mean that the astronaut has no weight! Weight is the force of gravity exerted on an object. As long as one is near to the planet (and being in orbit is still near) there is a force of gravity acting. For a further discussion of what is meant by weightlessness in orbit, see chapter 4.

[10] The force of gravity on an astronaut orbiting the Earth is slightly less than it was on the ground—after all he is now further from the centre of the Earth. However, it is not zero—weightlessness means the apparent absence of weight, not that gravity has vanished.

[11] However, on 7 February 1984 one of the space shuttle astronauts, Bruce McCandless, first tested a rocket-powered backpack that allowed him to move about without being connected to the shuttle.

[12] The balloon would have to be made of quite robust material in order to survive the pressure difference between the gas inside and the vacuum of space outside.

Intermission 1

The Saturn V booster rocket

At the time of its construction, the Saturn V was the most powerful rocket ever assembled by America. More powerful boosters were being tested by the Soviets, but they had an alarming tendency to explode. The Saturn V flew thirteen times between 1967 and 1973—including test flights—and still holds the record as being the only rocket design that has never exploded in flight.

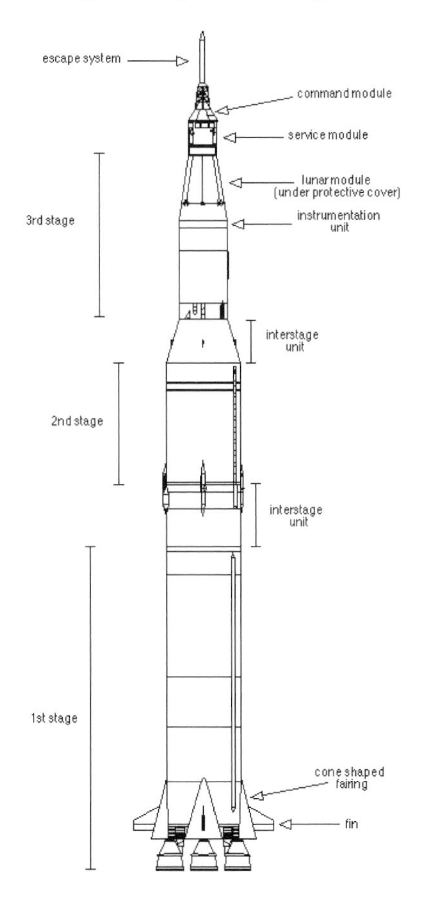

Standing 111 metres from base to tip, the Saturn V was an enormous structure. It broke down into several sections. There were three stages that contained propellant tanks and motors, two inter-stage sections that joined the stages together and which contained small rocket motors for in-flight manoeuvres, an instrument section that contained the control electronics for the whole rocket, and finally the Apollo spacecraft. Standing on top of the command module during launch was the emergency escape system. If a problem developed during the launch the rocket motors in the escape tower would fire, lifting the whole command module well clear of the rocket. The command module would then descend to the sea on its parachutes. It must have been a sobering moment for Alan Shepard (the first American in space) when as commander of Apollo 14 he entered the command module—the escape tower that would be his life-line in case of an emergency during launch was more powerful than the whole rocket that first lifted him into space!

The fairings at the bottom of the first stage served to smooth the flow of air over the engines in flight. They also contained the retro-rockets that fired to slow the spent stage once the second stage had taken over. The fins were another aerodynamic aid for the rocket during first-stage flight.

The design of the Saturn V was developed by the members of the Marshall Space Flight Center under Wernher Von Braun. The Saturn project had a history of different configurations, of which the Saturn I and Saturn Ib actually flew, testing design features that eventually went into the Saturn V (see appendix 3). Although various contractors were brought in to help with the detailed design and construction of the three stages, overall supervision and testing was carried out under the supervision of the Marshall team.

One of the reasons why Saturn V achieved its remarkable reliability was the testing and design procedure that was developed. A great deal of work went into specifying the 'environment' in which each part was supposed to work. This included information on the accelerations, stresses, loads, vibrations, temperatures, pressures, humidity and fatigues that it would be subjected to. The parts were then tested in situations more strenuous than their defined environment. Each part was required to demonstrate a reliability of 99.999 98%. Such an incredible figure is required as the overall reliability of the system is the product of the reliability factors of all the component parts. The design criterion called for an overall reliability of 99%. Another way of increasing reliability was to use redundant parts where possible.

Assembly of the rocket was an enormous enterprise in its own right. NASA constructed the giant Vehicle Assembly Building (VAB) in which the rocket could stand on top of a caterpillar tracked crawler that then carried it to the launch pad. The VAB stands 160 m high and is one of the largest volumes enclosed under a single roof in the world.

During the Mercury and Gemini programmes rockets were assembled, tested and fuelled on the launch pad, a system that had significant disadvantages. The rocket was exposed to the vagaries of the weather, and the launch pad was blocked for long periods of time. When it came to developing a new launch complex for the Apollo missions, the decision was taken to pioneer a new method—the mobile launcher. A pair of permanent launch pads (39A and 39B) were constructed along with three mobile launchers on which rockets were transported to the pads.

Manufacturers delivered the Saturn V components to the VAB where giant cranes, capable of lifting 200 tonnes, were used to stack them vertically onto one of the mobile launch pads. [It is said that the crane operators are capable of lowering a 40 tonne weight onto an egg without cracking its shell.] The mobile

Figure I1.1 A fully assembled Saturn V rolls out of the VAB on the crawler. Remember that the Saturn V itself stands 111 metres tall!

launchers were two-storey steel structures—7.6 m high, 49 m long and 47 m wide. While standing in the VAB they rested on six 6.7 m high steel pedestals. An umbilical tower 121 m high stood at one end of the launcher's base. Its job was to support the nine hydraulically activated service arms which carried the various propellant and power lines to the spacecraft as well as providing access to the stages for technicians. The crew entered the command module through a 'white room' at the end of an umbilical arm. Once at one of the pads, these umbilical arms were connected to the propellant storage tanks and generators sited near the pad. Some of the arms were moved to one side in the minutes before launch, the rest automatically swung away as the rocket lifted off.

The rocket itself sat over a 14 m square hole in the launcher through which the engine exhaust vented into a large trench on the launch pad.

Once the rocket had been assembled, checked out and the Apollo spacecraft installed, the next stage was to transport the whole stack to the launch site.

Figure I1.2 Top, a Saturn V first stage is lifted and winched over to be placed on a mobile launch pad. Bottom, the third stage of a Saturn V is lowered into place on top of a second stage. Support gantries on either side give technicians access to the rocket during assembly.

Figure I1.2 (*Continued*) Top, a completed lunar module is lowered into place. Bottom, an entire Apollo spacecraft stack is lifted for mounting on top of the third stage—note the LM legs visible at the base of the stack.

Figure I1.3 The Apollo 8 Saturn V sits on pad 39. The pipe work up the side of the service tower is visible on the left. On the right is a service gantry that can be rolled away as launch approaches.

Various methods of doing this were considered by the NASA engineers, including using a large barge (the first and second stages were transported to Kennedy in this manner). Eventually a crawler was designed and constructed by the Marion Power Shovel Company. This giant machine carried the full rocket stack and launcher the 5.6 km journey to pad 39 at 1.6 kilometres an hour. Riding on eight caterpillar tracks, each one being 2 m wide and 12.5 m long, the crawler was equipped with a sensing system that kept the rocket within 1 degree of being vertical. A set of 47 m long pipes ran in a giant X under the top plate, connected to hydraulic jacks at each corner. If the top moved by as little as 1 cm the jacks could level the system again. This was especially important during the climb up the 5° slope to the launch site.

Figure I1.4 The launch complex of pad 39A. There are several interesting features to be seen in this picture. In the middle distance is the Vehicle Assembly Building at the far end of the crawler way, which looks like a two-lane road. The crawler itself (wide enough to span both lanes) can be seen proceeding back to the VAB just to the right of where the road turns. A Saturn V sits on its mobile launcher at complex 39A. There is a slope up to the pad so that it sits above ground level, allowing the flame trenches to run underneath.

The crawler drove into the VAB and under the mobile launcher. The jacks rose to lift the launcher slightly above the pedestals on which it was sitting, allowing the crawler to drive out again with the launcher on top. The reverse process was used to install the launcher on pedestals at the pad.

Each of the Apollo launch pads was 0.65 square kilometres in size and constructed of heavily reinforced concrete. The propellant needed for the various stages of the rocket was stored at the pad along with the pumping equipment needed to transfer the liquids to the stages prior to launch. The mobile launcher sat on pedestals over a 17.7 m wide flame trench.

At the time of launch a 590 tonne flame deflector was moved under the mobile launcher on rails and raised by hydraulic rams into position. Its purpose was to absorb the blast of the exhaust and to deflect the flame along the line of the trench (without this there was a danger that the Saturn V could be damaged by its own exhaust reflected back). It was covered with 11 cm of refractory concrete made from a combination of volcanic ash and calcium aluminate (acting to bind the ash particles together). During launch 2 cm of the material was worn away.

The combination of pad, mobile launcher and flame trench was cooled with a water deluge system.

The deck of the mobile launcher was soaked by 29 water nozzles for the first 30 seconds after launch at a volume rate of 189 000 litres per minute. After that the volume rate was reduced to 76 000 litres per minute for a further 30 seconds. The flame trench was cooled at 30 000 litres per minute, starting 10 seconds before lift off.

Stage details

(Taken from the press kit issued prior to the Apollo 11 launch.)

First stage

Height: 42.06 m

Largest diameter: 10.06 m

Weight fully fuelled: 2 278 247 kg

Weight empty: 130 975 kg

5 Rocketdyne F-1 engines burning RP-1/LOX propellant producing a total thrust of 33 375 000 N, uprated over a period of time to a final thrust of 35 155 000 N. This stage burnt for 170 seconds, carrying the rocket to an altitude of about 61 km and a speed of 9654 km h^{-1}.

Inter-stage unit mass: 4583 kg.

Second stage

Height: 24.84 m

Largest diameter: 10.06 m

Weight fully fuelled: 480 432 kg

Weight empty: 36 250 kg

5 Rocketdyne J-2 engines burning LH_2/LOX propellant producing a total thrust of 4 450 000 N. The second stage burnt for 395 seconds to an altitude of about 184 km and a speed of 24 617 km h^{-1}.

Inter-stage unit mass: 3665 kg.

Third stage

Height: 17.77 m
Largest diameter: 6.61 m
Weight fully fuelled: 118 171 kg
Weight empty: 11 340 kg

1 Rocketdyne J-2 engine burning LH_2/LOX propellant producing 890 000 N of thrust. The third stage's initial burn to place Apollo in Earth orbit lasted for 165 seconds. At this point the spacecraft was at an altitude of 185 km with a speed of 24 617 km h^{-1}. The third stage was lit again to place Apollo onto an orbit that would carry it out to the Moon. This trans-lunar injection burn lasted 312 seconds and boosted the spacecraft's speed to 39 420 km h^{-1}.

Instrument unit mass: 1953 kg.

Total Saturn V mass when fuelled: 2 887 051 kg.

Payload capacity

(The total mass of components above the instrument unit.)

Apollo 8's Saturn V: 36 000 kg
Apollo 17's Saturn V: 53 000 kg

These two figures show the soundness of the Saturn V design and how the engineers were able to increase the thrust produced by the various stages.

Contractors for the Saturn V

First stage	Boeing Company
Second stage	North American Aviation (Rockwell International)
Third stage	Douglas Aircraft Company (McDonnell Douglas)
F-1 engines	Rocketdyne
J-2 engines	Rocketdyne
Instrument unit	IBM/Marshall Space Flight Center

Chapter 3

Rocketry

In this chapter we will look at the physics of multistage rockets and the principal design features of the Saturn V. We will also take a look into the future to see how rocketry might develop in the new millennium.

3.1 Faltering starts

Towards the close of World War II the German rocket scientist Werner Von Braun brought his team of engineers together in secret to discuss what they would do after the war. The Nazis had forced them to work on developing rockets (such as the V2) that they hoped would provide a last-minute weapon to defeat the allies. Von Braun's team dreamed of using this technology to explore space[1]. They decided that their best chance of pursuing this dream was to surrender to the United States (they did not fancy going to Russia—they had seen enough of totalitarian regimes!). Meanwhile the Americans were running *Operation Paperclip*—combing occupied Germany for scientists who might be useful after the war. The Nazis instructed Von Braun to destroy his papers (he hid them in an abandoned mine) and shipped the team away for what they believed would be execution. Fortunately, in the confusion of the last days of the war, they managed to surrender to the American forces.

However, when they arrived in America they found a state that was not as keen to develop space rockets as they had hoped. The war had ended and Congress was too concerned with other matters to vote money into rocket and space research. Von Braun and his team were dumped at Fort Bliss in Texas to tinker with captured V2s and teach rocketry to interested members of the Army. This unfortunate state of affairs continued for six years. During that time the Soviets were putting their captured Germans to far more 'constructive' use—developing ballistic missiles.

Eventually in 1950, the Army became convinced that Soviet rocket research was taking place and mobilized Von Braun. His team was moved to Huntsville, Alabama, were they worked at Redstone Arsenal (where previously the Army had loaded explosives onto shells and bombs—it was closed after the war) to start the development of missiles. Once again they were being forced to construct weapons rather than peaceful rockets. Just weeks after they arrived, North Korea invaded the South and the word came down from President Truman—develop a ballistic missile capable of firing a nuclear warhead 200 miles.

Three years later the first *Redstone* missile was launched from the military test range at Cape Canaveral in Florida.

By 1956 the Air Force was developing the *Thor*, *Atlas* and *Titan* missiles (the latter pair capable of firing warheads 5000 miles) and the Army was working on *Jupiter*. However, none of them were flight ready and there was a need to test the warhead design's ability to re-enter the Earth's atmosphere. Von Braun was asked to further develop the *Redstone* to carry out these tests. This they did by lengthening the rocket, modifying the engines and adding two upper stages. The modified rocket was called *Juno-C*. It worked perfectly on the first test. Von Braun was quite aware that a small further stage added to the top would be sufficient to launch a satellite into orbit. They developed the necessary equipment and asked for permission to proceed. The response from Washington was rather negative. Rumour was that the President was not about to allow the first satellite to be launched by ex-German scientists. Two years earlier he had sanctioned the development of a non-military booster, *Vanguard*, to put the first satellite into orbit. Von Braun was worried. He realized that the Soviets were just as capable of adapting a ballistic missile to carry satellites as he was. He was convinced that *Vanguard* would not work in time.

Von Braun was proven correct. On 4 October 1957 the Soviets placed *Sputnik* in orbit. The modified *Juno-C* was sitting in a shed.

Thirty days later the Soviets launched a much bigger satellite (500 kg) with a live dog on board.

Vanguard's first attempt, under enormous pressure from Washington, took place on 6 December that year. The rocket was shipped to Cape Canaveral where, with extensive press coverage, it attempted to launch a 14 kg satellite.

The launch was certainly spectacular. One of the commentators watching exclaimed, 'it was so quick, I really did not see it'. In fact, *Vanguard* had exploded on the pad, blasting the satellite into the air to land in the scrub surrounding the pad. Once there it dutifully started transmitting its radio signals.

Another of the assembled press corps simply said 'why doesn't someone go out there, find it, and kill it?'

Finally, President Eisenhower relented and gave Von Braun's team the chance. The modified *Juno-C* was taken out of storage, further changes were made, and on 31 January 1958 it successfully placed the *Explorer 1* satellite in orbit first time. On board was a Geiger counter developed by Dr James Van Allen, who consequently made the first scientific discovery of the new space age—that the Earth is surrounded by regions of radiation now known as the *Van Allen belts*.

Von Braun's *Redstone* would soon be used to launch the first American into space—but not before further political indecision had allowed the Soviets to place the first man in orbit. The road that led to the development of the Saturn V by Von Braun would be long and difficult, but much of the basic rocketry needed to place a man on the Moon was developed in those early years.

3.2 Thrust

Thrust is the rocket engineer's term for the force applied by the propulsion gases pushing against the rocket. Applying Newton's second and third laws to the case of a rocket engine produces the following expression for the thrust:

$$\text{thrust (N)} = \text{exhaust velocity (m s}^{-1}) \times \text{propellant consumption rate (kg s}^{-1})$$

$$T = u \times \frac{\Delta m}{\Delta t}$$

where u is the exhaust velocity.

Mathematicians use Δ to mean 'change in', so that Δm means the change in mass (not some quantity $\Delta \times m$!), Δt is the change in time—or the amount of time elapsed. If v is the velocity of the rocket at any moment, then $\Delta v / \Delta t$ is the rate at which the velocity is changing, or the acceleration.

From this expression we can see that a large thrust will be produced if the exhaust gases move at the highest speed possible and propellant is consumed at a rapid rate.

Some idea of the physics behind this can be gained by noticing that the exhaust gas speed multiplied by the change in the propellant mass (i.e. the mass that has been ejected as exhaust) is the momentum of the exhaust. So, the expression for the thrust is related to the momentum of the propellant ejected as exhaust per second:

$$T = u \times \frac{\Delta m}{\Delta t} = \frac{u \times \Delta m}{\Delta t} = \frac{\Delta(mu)}{\Delta t}.$$

The last part of this expression is saying that the thrust (which remember is the same as the force) is the change in momentum per second. This is another way of writing Newton's second law of motion and corresponds to the more general manner in which he devised it. In Newton's full form the second law of motion reads:

$$\text{force applied} = \frac{\text{change in momentum}}{\text{time}}.$$

Later in this chapter we will find that this is a more useful way of thinking about the second law when dealing with rockets.

3.3 Propellant

A key factor in producing the highest possible thrust is the choice of propellant to be used. There are many different chemical combinations that have been developed for use in rocket motors, all of which have specific advantages and disadvantages. In every case two chemicals are mixed to produce the reaction. One is acting as the fuel—the Saturn V first stage used RP-1 (kerosene)—and the other as an oxidizer—liquid oxygen in the case of the Saturn V.

When you light a fire you are triggering a chemical reaction. The flames are gases released from the burning materials. Combustion will only take place once a certain temperature has been reached. However, once it is triggered the energy produced by the reaction itself will maintain the temperature until the reactants are exhausted. It is easy to forget that one of the chemicals involved in the reaction is the oxygen in the air. Often the reason that fires go out is oxygen starvation (this is why fire fighters sometimes smother fires in foam—it blocks off the oxygen).

With a rocket, the oxidizer has to be provided. The volume of chemicals being reacted per second makes it wholly impractical to rely on oxygen within the air. Not only that, but as the rocket climbs higher into the atmosphere the density of oxygen present drops rapidly.

Each possible fuel/oxidizer combination can be given a rating, which relates to the amount of thrust produced by the combination. Rocket propellants are rated according to their *specific impulse*, I_{sp}, which is the amount of time during which you can completely burn 1 kg of propellant to provide 10 N of thrust. The larger the specific impulse the longer 1 kg of propellant will provide thrust and so the better the propellant. From the previous section, thrust is given by:

$$T = u \times \frac{\Delta m}{\Delta t}$$

so to provide 10 N of thrust, 1 kg of the propellant must be burnt in a time Δt where:

$$\Delta t = \frac{u\Delta m}{T} = \frac{u \times 1\ \text{kg}}{10\ \text{N}} = I_{sp}.$$

Now, we can relate 10 N to 1 kg as on Earth 10 N is the weight of 1 kg, in other words 10 N $= 1$ kg $\times\, g$ with g being the strength of gravity on the surface of the Earth. Plugging this fact into our expression for Δt gives:

$$I_{sp} = \frac{u \times 1\ \text{kg}}{10\ \text{N}} = \frac{u \times 1\ \text{kg}}{1\ \text{kg} \times g} = \frac{u}{g}.$$

Consequently the specific impulse is the ratio of the exhaust speed to the strength of gravity on Earth. Another, equivalent, way of defining the specific impulse is to say that it is the thrust produced divided by the weight of propellant used per second:

$$\begin{aligned} I_{sp} &= \frac{\text{thrust}}{\text{weight of propellant used per second}} \\ &= \frac{u \times \Delta m/\Delta t}{\Delta W/\Delta t} = \frac{u \times \Delta m/\Delta t}{g \times \Delta m/\Delta t} = \frac{u}{g}. \end{aligned}$$

A variety of different propellant mixtures are compared in table 3.1. The choice of propellant is not simply down to picking the combination with the largest specific impulse. The difficulty in storing the material and the context in which it is to be used must be taken into account as well. For example, hydrogen and oxygen can only remain in liquid form provided the temperature remains near to absolute zero—which gives storage problems.

The Saturn V used RP-1/LOX in the first stage and LOX/LH_2 in the second and third stages. Although the LOX/LH_2 combination has the higher specific

Table 3.1 Propellant mixtures.

Fuel	Oxidizer	I_{sp} (s)	Comments
RP-1	liquid oxygen (LOX)	303	LOX needs to be stored at cryogenic temperatures
liquid hydrogen (LH_2)	LOX	453	both LOX and LH_2 need to be stored at cryogenic temperatures
aerozine 50 (similar to kerosene)	nitrogen tetroxide	320	easily stored, comparatively cheap
hydrazine	nitrogen tetroxide	300	corrosive, hypergolic
aluminium polymer	ammonium perchlorate	266	solid propellant

impulse, it was not used in the first stage as a high-efficiency multiple-stage rocket design demands that the first stage carries the bulk of the propellant for the whole rocket (see later in this chapter). It was decided that storing that much LOX and LH_2 in the propellant tanks, both of which would have to be kept ultra-cold to prevent the propellant boiling into a useless gas, was too difficult. Furthermore, to store the required mass of LH_2 in the first stage would have required impracticably large tanks (hydrogen is only 1/12th of the density of RP-1, so that an equivalent mass requires 12 times the volume). The higher I_{sp} fuel was used in the later stages as the mass of propellant required was much less.

Pictures of the rocket taken during lift off show clouds of white material falling off its sides. This is ice that has formed on the outer shell of the propellant tanks. Up to 630 kg of water vapour in the air outside froze onto the metal surface. Most of it fell off in the first seconds after launch[2].

In some situations, for example a satellite changing orbit, or the service and lunar module engines used in Apollo, a very reliable mixture is required. The lunar module ascent stage motor had to fire the first time or the explorers would have been stranded on the Moon. In such key circumstances, it is better to design an engine with the simplest mode of operation and so the smallest number of parts that can go wrong. Hypergolic propellant mixtures are ideal for this. Such propellants spontaneously ignite when mixed, unlike RP-1/LOX, for example, which requires an ignition source. All a hypergolic engine has to do is pump the chemicals from their storage tanks into a thrust chamber and the engine starts automatically.

A particularly simple form of rocket motor uses solid propellant. This is a mixture of both fuel and oxidizer in a semi-solid, putty-like form. Typically this putty is packed round the inside of a cylindrical engine body with a hole left in the middle running down the length of the engine. An igniter placed at the top sends a burst of flame down the length of the hole. This ignites the putty at every point down the length of the engine. Thrust is produced by the hot gases pushing up against the top of the cylinder, with the exhaust gases shooting down the length of the hole and out of the bottom. The amount of thrust produced depends on how much of the propellant is burning at any one time. If the hole down the centre has a simple circular cross section, then as the putty burns away increasingly large surface areas of new putty are exposed. In such a simple design the thrust would increase with time. Rocket designers can produce very cleverly shaped holes in the putty, which allow some control of the variation of thrust with time to be designed in.

This is an extremely simple design of rocket and the propellant used has the considerable advantage of being easy to handle and not requiring cryogenic temperatures. However, as the fuel tank and the combustion chamber are essentially the same, the rocket body has to be made very strong, and so heavy[3]. Another disadvantage of solid rockets is that once they have been ignited they cannot be stopped.

However, their simplicity makes them ideal for one special application. Solid rockets are used as additional engines for the space shuttle during lift off. Once their propellant has been exhausted, the solid rockets are jettisoned and parachute into the sea, from where they are recovered, refilled with propellant and used again on a later launch.

3.4 Applying Newton's laws to a spacecraft

In order to produce the thrust required to lift the Apollo spacecraft off the ground, the engines of the Saturn V first stage consumed propellant at an average rate of nearly 12 tonnes per second. Once the propellant had been used up, some 170 seconds into the flight, the rocket was moving at a speed of 9600 km h^{-1} (kph) (6000 mph).

Now at first glance these figures do not add up. From stationary to 9600 km h^{-1} in 170 s is an average acceleration of 15.7 m s^{-2}—a lot greater than the acceleration that we quoted in the previous chapter (1.92 m s^{-2}).

As the rocket burns its mass decreases. The propellant is consumed and ejected out of the engines as exhaust. Even with a constant force being applied by the engines, the acceleration of the rocket will increase with time as its mass is steadily decreasing. The acceleration that we calculated was for a fully fuelled rocket at the moment of lift off. As the engines are consuming propellant at the rate of 12 tonnes a second, the mass of the craft drops rapidly and so the acceleration increases at some rate.

In such circumstances, Newton's second law as quoted in the previous chapter cannot be used to calculate the acceleration. The form:

$$\text{force} = \text{mass} \times \text{acceleration}$$

only works if the mass of the accelerating object is not changing. In order to deal with the more general case, the more sophisticated form mentioned earlier in this chapter has to be used:

$$\text{force applied} = \frac{\text{change in momentum}}{\text{time}}.$$

This form is more general as it reduces to the previous equation in the special case of constant mass. After all:

$$\text{momentum} = \text{mass} \times \text{velocity}$$

so

$$\begin{aligned}\text{force applied} &= \frac{\text{change in (mass} \times \text{velocity)}}{\text{time}}\\ &= \text{mass} \times \frac{\text{change in velocity}}{\text{time}}\\ &= \text{mass} \times \text{acceleration.}\end{aligned}$$

When the mass of the object is changing the relationship becomes:

$$\begin{aligned}\text{force applied} &= \frac{\text{change in (mass} \times \text{velocity)}}{\text{time}}\\ &= \text{mass} \times \frac{\text{change in velocity}}{\text{time}} + \text{velocity} \times \frac{\text{change in mass}}{\text{time}}\end{aligned}$$

which is a very complicated looking expression. We can make it look slightly friendlier by using the standard abbreviations:

$$\text{force applied} = \frac{\text{change in (mass} \times \text{velocity)}}{\text{time}} = \frac{\Delta(mv)}{\Delta t} = m\frac{\Delta v}{\Delta t} = v\frac{\Delta m}{\Delta t}.$$

However we look at it, the expression is difficult to use as the various terms are not always constant. Newton had to invent a new brand of mathematics in order to deal with such situations—calculus.

For our purposes the important matter is the final expression that can be derived by applying this to a rocket in flight (although those with a mathematical turn of mind can check the result in appendix 4). There are two possible situations to consider—launch from the Earth's surface, when gravitational pull needs to be taken into account, and manoeuvring in deep space when the effects of gravity can be ignored.

No gravitational force

The change in speed (ΔV, pronounced 'delta-V') that a rocket of mass M_R (the rocket's casing[4] empty of propellant + payload) can achieve by burning propellant of mass M_P in the absence of a gravitational force is:

$$\Delta V = u \times \ln\left(1 + \frac{M_P}{M_R}\right)$$

u being the speed at which the exhaust gases are ejected out of the back of the rocket. 'ln' is the natural logarithm—a well known mathematical function.

We can rewrite the equation slightly to include the specific impulse by using the fact that $I_{sp} = u/g$:

$$\Delta V = g \times I_{sp} \times \ln\left(1 + \frac{M_P}{M_R}\right).$$

Delta-V is the currency of rocketry. All manoeuvres that need to take place in space require velocity changes (delta-Vs) and this equation specifies the amount of propellant needed to carry out the manoeuvre. Given that a spacecraft carries a limited amount of propellant, every change of orbit or docking manoeuvre or speed change has to be carefully budgeted for. The delta-V budget of a complete mission has to be worked out in advance, with a small amount of latitude for emergencies. Every kilogram of propellant that is carried on the spacecraft is extra mass that could have been used as payload. Mission designers are very thorough in eliminating any excess weight from the spacecraft.

It is worth noting that the time taken to achieve this final velocity is not part of the expression. This is only true in deep space when the gravitational force can be neglected. In this case, if the propellant is burnt slowly, the rocket will take a long time to achieve its final speed. On the other hand, if the propellant is burnt quickly, then that speed will be achieved in a much smaller period of time, but the final speed will be the same in both cases.

The ratio between the mass of propellant burnt and mass of the rocket (casing + payload) is critical. In order to achieve a high speed the mass of propellant burnt must be much greater than the mass of the rocket.

Figure 3.1 shows how the final velocity of a hypothetical rocket with an exhaust speed of 3 km s^{-1} varies with mass ratio. The curve shows that a severe penalty has to be paid for increasing the final speed. Doubling the speed from 4 km s^{-1} to 8 km s^{-1} implies that the mass ratio must increase from just under 3 to about 13. Assuming the mass of the empty rocket remains the same, this means that the mass of propellant has to increase by 13/3 or 4.3. Of course, it is not possible to increase the mass of propellant without increasing the mass of the empty rocket as the propellant has to be stored in larger tanks. In any case even more propellant would have to be carried with the rocket in order to slow it down again at the end of the journey.

Rocket speed is worth far more than its weight in propellant.

Ratios very much greater than 10:1 are impractical as the tremendous weight of the propellant has to be supported by robust tanks, which adds to the weight of

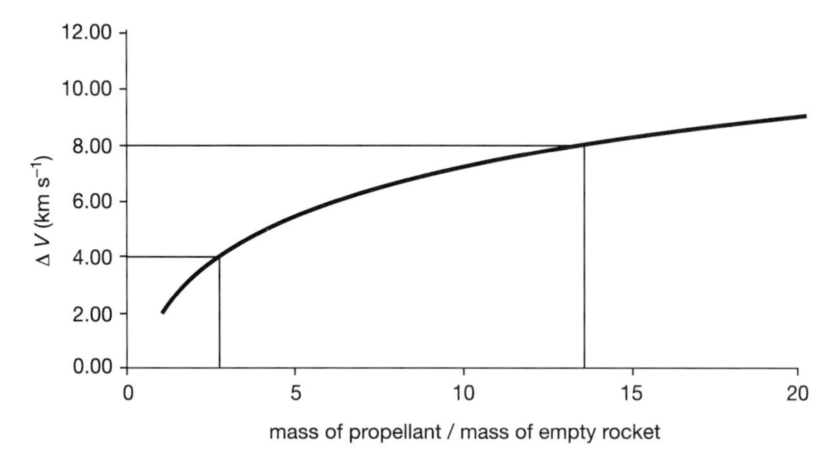

Figure 3.1 How the final speed of a rocket varies with the mass ratio.

the rocket so reducing the ratio. Exhaust speeds much greater than 3 km s^{-1} (which after all is 9 times the typical speed of sound at sea level) are difficult to achieve. The Saturn V's first stage produced an exhaust velocity of 2.7 km s^{-1} and had an excellent mass ratio of 15:1.

The effect of gravitational pull—launch

The version of the delta-V equation that we have been using so far is only valid if there is no gravitational force pulling the spacecraft. While this may be a reasonable approximation in deep space, it is certainly not true during launch, and even during the flight of Apollo (which had both the Earth's and the Moon's gravity acting on it).

Fortunately the modification needed to the equation is minor, although the effect of this modification is quite important. In the presence of a gravitational field of strength g, the delta-V achieved by burning propellant of mass M_P is:

$$\Delta V = g \times I_{sp} \times \ln\left(1 + \frac{M_P}{M_R}\right) - gt.$$

The most instantly significant aspect of this new expression is that time has been introduced. Up to now I have emphasized that the delta-V achieved is independent of the rate at which the propellant is burnt. In the presence of gravity, however, this is not true. The longer it takes to burn the propellant, the smaller the delta-V. This is why it is vital to pump propellant into the engines at a huge rate during launch.

One way of making this more intuitively obvious is to consider that the thrust achieved by the engines must be greater than the weight of the rocket, or it will never leave the ground. The greater the thrust the faster the rocket will accelerate—or in other words achieve a greater delta-V in a shorter time. In order to increase the thrust, propellant must be used at a rapid rate and we are back to the same conclusion again.

3.5 Real rocket engines

Translating the basic physics of how rocket motors work into a useable engine is a complex engineering task. Many different components have to be optimized in order for the engine to deliver the maximum possible thrust from the propellant that it is using. The Saturn V used two different designs of engine.

The F-1 running from RP-1/LOX powered the first stage and the other stages used the J-2 design running from LH_2/LOX. Without going into the detailed engineering of these different designs, there are some interesting features that are common to all engines that are worth discussing. Figure 3.3 is meant to represent the common features in a rocket engine design. It is not a faithful rendition of either the F-1 or J-2 engines.

Propellant delivery

One of the primary goals in engine design is the delivery of fuel and oxidizer to the thrust chamber in the right proportions and at the right rate. In the case of the F-1 fuel was delivered at 58 000 litres per minute and oxidizer at 93 900 litres per minute. Both liquids fed into the thrust chamber via pumps which were driven off a common shaft from a gas turbine. The high-velocity gas required to drive this turbine was derived from a gas generator in which small amounts of fuel and oxidizer were mixed. At the moment of engine ignition, valves opened to allow pressurized fuel and oxidizer into the gas generator where they were ignited by small explosives. The reaction produced the exhaust gases which passed through the turbine blades at 77 kg per second. The turbine, generating 41 MW, drove the pumps which channelled more propellant to the gas generator as well as to the thrust chamber.

It was important to try to maintain a constant pressure in both the fuel and oxidizer tanks during the flight. If a partial vacuum was allowed to develop above the liquids (as it does whenever you drain a liquid from a sealed container), then the pumps would have a harder time drawing liquid out of the tanks. As the levels dropped, evaporated liquid would partially fill the

Figure 3.2 One of the five F-1 rocket engines (top) that powered the first stage (bottom) of the Saturn V.

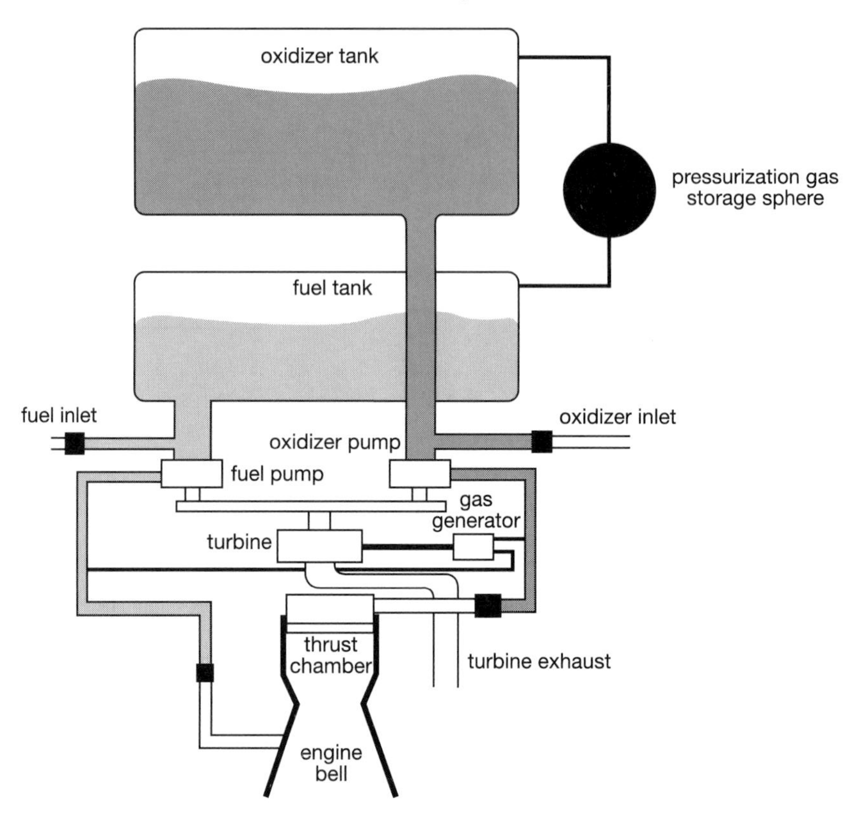

Figure 3.3 Schematic of a basic engine design. The fuel reaches the thrust chamber by circulating in pipes round the outside of the engine bell—it therefore acts as a coolant for the engine bell at the same time.

empty space, but there would still be a pressure drop in the engine feed lines. In order to try and maintain a constant pressure a gas that did not react with the content of the tank was pumped in over the liquid. A similar problem would develop in flight at the moment of staging. Once a stage's engines had shut down and before the next stage had lit up, the rocket would be freely falling under gravity (although of course it would still be going upward!). Under these circumstances the propellant tends to float in the tanks, making it difficult to pump. To counter this, small solid rocket 'ullage' motors would fire in the next stage prior to the main motors. This would give the stage an upward acceleration, settling the liquid before the pumps cut in at main engine ignition.

Both the thrust chamber and the engine bell must be cooled in operation in order to prevent them melting under the extreme temperatures of the exhaust

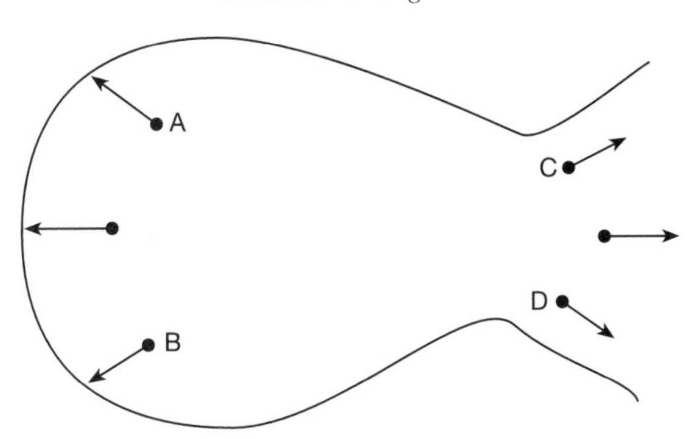

Figure 3.4 Thrust in a balloon.

gases. The general principle is to circulate cooler liquids around the outside. When a cool liquid passes over a hot body it will conduct energy away, reducing the temperature of the body. The F-1 engine's thrust chamber was kept cool by circulating the cryogenic fuel around the chamber before it entered to react with the oxidizer. The engine bell was kept cool by routing the turbine exhaust around the structure.

Nozzle design

At the far end of the thrust chamber the exhaust gases emerge into the surroundings via the *nozzle*, the purpose of which is to convert as much of the random motion (pressure) in the exhaust gas into concerted motion (flow) as possible.

The balloon mentioned in the previous chapter does not provide thrust very efficiently as there is no control over the direction in which the molecules emerge from the hole in the skin. If the best use is to be made of the energy in the molecules, then they should all emerge parallel to the direction of flight. Remember that the thrust is actually due to the molecules that are striking the skin of the balloon on the opposite side (or in the case of a real engine, the exhaust molecules striking the top of the thrust chamber). The random nature of the molecules' motions in both cases ensures that the molecules providing the thrust are striking the balloon skin (thrust chamber) in all directions. In many cases the forces are cancelling each other out (see figure 3.4).

In figure 3.4 molecules A and B exert forces on the balloon's skin, but they do not propel the balloon forwards effectively. This is because they are striking the balloon at an angle to the direction of flight. Part of the impact force of molecule A will push the balloon upwards; part of the impact force of molecule B will push the balloon downwards. Not only is this not the direction of flight, but the impacts of the two molecules are partially offset against each other. Molecules C and D leaving the balloon carry some energy with them and ensure that the impacts of A and B do provide some net thrust to the balloon, but not as efficiently as the horizontal pair of molecules shown in the same diagram. If the hole in the balloon could be designed so that only molecules that were moving parallel to the direction of flight left the balloon, then the thrust produced would be more efficiently employed. One way to do this is to reduce the size of the hole. A small hole will only allow molecules out that arrive close to straight on. However, a natural consequence of this is that fewer molecules can get through. As the thrust is partially dependent on the rate at which mass is consumed (or ejected from the balloon), reducing the size of the hole also reduces the thrust.

A real engine has to be more cleverly designed than this to ensure that as much mass as possible is allowed out of the engine per second, and that the molecules are leaving parallel to the direction of flight and moving as rapidly as can be arranged.

A gas exerts pressure because its molecules are moving about rapidly and in random directions. As a result they collide with each other and with the walls of the container. These collisions exert forces on the walls which translate into the pressure the gas exerts.

A gas will flow when the molecules have a concerted drift speed in a given direction in addition to their rapid random motion. Consider the atmosphere in the Apollo command module. The gases inside the capsule will be exerting a pressure due to their random motion. However, the molecules must also be sharing in the forward motion of the capsule, or they would be left behind as it travels along.

At the top end of the thrust chamber the gas is not moving very fast, but exerting a high pressure. At the bottom of the engine bell the gas is moving quickly (flowing), but the pressure has dropped to being equal to the external pressure.

The nozzle acts to convert the gas pressure at the throat to a directed flow of exhaust out of the engine bell. This has two effects. Firstly, if the consumed propellant leaves the thrust chamber as exhaust at the greatest mass rate per second, then more can be pumped in to continue the reaction. Secondly the process of converting pressure into high speed flow causes a back reaction that maximizes the pressure at the top of the thrust chamber. It is this pressure, after all, that is pushing the rocket forward.

The most commonly used design is the de Laval nozzle, after Dr Carl de Laval (1845–1913) the Swedish engineer who first conceived the idea, which consists

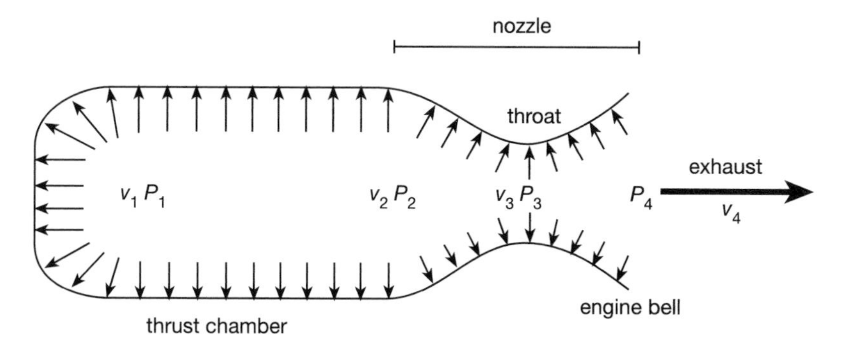

Figure 3.5 A convergent–divergent nozzle design for a rocket engine. The arrows represent the pressure exerted by the gas at that point. On the left the flow velocity of the exhaust gas, v_1, is very low, and on the right v_4 is hypersonic.

of a gently tapering exit from the thrust chamber down to a minimum diameter (the throat) and then an expanding area (the engine bell) leading to the surroundings.

The tapering exit hole leading to the throat increases the flow of gas leaving the thrust chamber in the same way that putting your finger over the end of a hosepipe will increase the speed at which the water is escaping.

In fact, two things are happening. Firstly, the tapering exit hole is preferentially selecting molecules that are travelling parallel to the centre line of the engine. Secondly, molecules that are bouncing back off the sides are not escaping: they are colliding with more molecules arriving at the far end of the chamber. This has the effect of building up the density of gas behind the throat. Consequently the pressure, P_2, is increasing there (and back through the chamber, i.e. P_1). As a result of this pressure build up there are more collisions acting on molecules pushing them down the throat than there are pushing them back, so there is a net acceleration of molecules into the throat of the nozzle. The amount by which the molecules are accelerated depends on the pressure difference between the start of the nozzle and the throat (P_2 and P_3 on the diagram).

With more molecules travelling parallel to the sides of the throat, fewer of them are striking the sides and so the pressure of the gas on the side walls is reduced. This effect is quite commonly used in different engineering applications[5].

With a rocket engine the taper on the throat is designed to build the speed of gas flow up to the speed at which sound waves would travel through the exhaust gas. Technically this condition is known as having a *choked nozzle*[6]. Of course

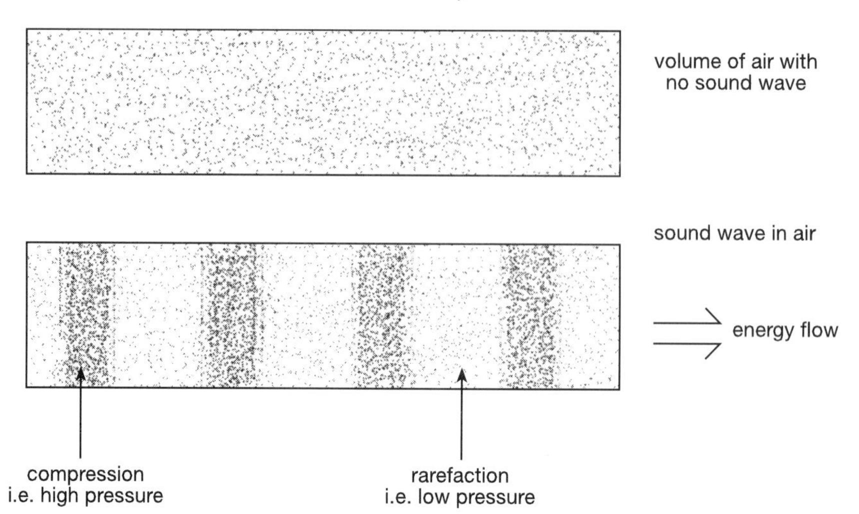

Figure 3.6 The upper diagram shows a pipe containing air. The dots represent molecules that are in rapid, random (thermal) motion. The molecules are travelling at great speed, but on average they are not going anywhere. The lower diagram shows the pipe with a sound wave travelling through it. Now the molecules cluster into high-pressure regions and low-pressure regions. A moment later, the high-pressure regions will have become low-pressure regions, and vice versa. This happens due to the collisions between the molecules. From this we can see that sound cannot travel at a speed that is faster than the molecules' thermal motion.

this does not mean that the throat is physically choked—exhaust gases can still travel down, driven by a pressure difference.

The speed at which a sound wave travels through a gas depends on the motion of the gas molecules. A sound wave is a sequence of *compressions* (high-pressure regions) and *rarefactions* (low-pressure regions) that move through the gas.

In a gas which is not flowing, a disturbance travels through due to the collisions between the molecules. The difference between a sound wave and a draught of wind is that in the former case the molecules do not move from their average positions, but in the latter there is a net drift of molecules along. This drift speed is superimposed on the random thermal motion of the molecules in the gas.

A sound wave cannot travel through a gas at a speed greater than the average speed of the molecules *due to their thermal motion*. However, there is no reason why the *drift speed* cannot be greater than the *thermal speed*. This is a curious situation, but there is nothing in physics to prevent it. When this happens, the gas is flowing at a speed greater than that at which sound can travel through the gas. The gas is in *hypersonic flow*.

A typical engine design will arrange for the ratio between P_2 and P_3 to be great enough to *apparently* accelerate the gas in the throat to hypersonic speeds. In fact, in the situation of a constricted throat such as this the gas can only be accelerated up to the speed of sound. Instead of going beyond this speed, the pressure difference compresses the gas. This is the state of the gas as it moves beyond the throat into the widening engine bell.

With the gas flowing through the throat at a speed greater than that of sound waves through the gas, the gas is no longer sensitive to pressure changes taking place beyond the end of the nozzle[7]. After all, a sound wave is a pressure wave communicated by the thermal motion and if the molecules are drifting forwards at speeds greater than their thermal motion there is no way for any pressure change to communicate back through the gas.

Any change in the pressure of the gas further down into the engine bell (P_4) has no effect on the speed of the molecules further back in the throat. Once the ratio of P_2 to P_3 is great enough for the gas to be accelerated to the speed of sound, any change in P_4 will have no effect on the speed in the throat.

Under normal circumstances, when a gas arrives at a widening aperture, such as the engine bell, the speed of the molecules would start to drop again. However, with a compressed sonic velocity gas a different and complex effect happens. The compressed gas expands explosively, releasing energy into the speed of the molecules[8]. The result of this is that the molecules are actually accelerated further and travel at hypersonic velocity out of the engine bell.

The exact shape of the engine bell has to be very carefully designed. The trick to obtaining the greatest efficiency of thrust is to match the pressure of the gas emerging from the end of the bell to the pressure of the surrounding atmosphere. Clearly, as a rocket travels higher the pressure of the atmosphere decreases (indeed in space it is zero), so that an engine bell cannot be optimized for every height. The F-1 and J-2 engines had different shaped nozzles as the stages that used them were designed to work at different altitudes.

Remember that the sideways pressure of the gas is due to molecules that are not perfectly aligned along the direction of the rocket's flight. If the external atmospheric pressure is greater than the sideways pressure of the exhaust gas as it leaves the engine bell, then collisions between the molecules in the atmosphere and those in the exhaust will tend to push the exhaust inwards. This decreases the efficiency of the thrust. On the other hand, if the exhaust gas pressure is greater than that of the atmosphere then there is a net tendency for molecules

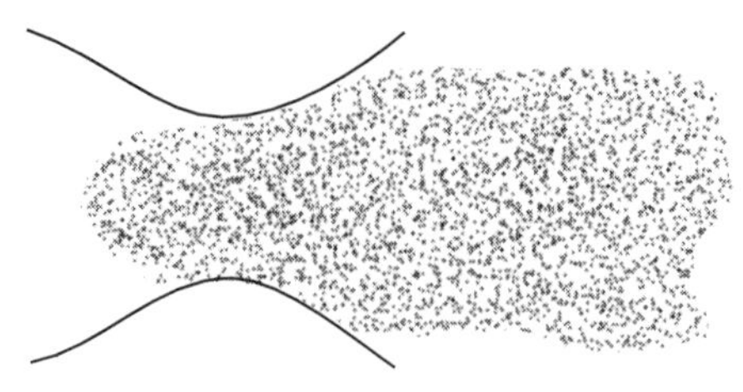

exhaust pressure equal to external pressure

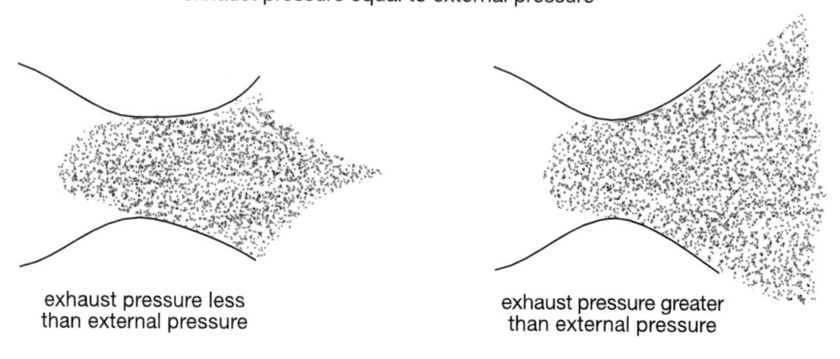

Figure 3.7 The effect of different external pressures on the shape of the exhaust plume from a rocket engine.

in the exhaust to move outwards (they will not be pushed back in by molecules in the atmosphere), again reducing efficiency.

There is another way of looking at this whole argument.

Figure 3.5 only shows the pressures acting within the system of the thrust chamber and nozzle. In the atmosphere, the pressure of the air is acting externally and equally over the whole shape of the rocket. Air molecules striking the front of the ship will tend to slow it down[9]. Air molecules striking exhaust molecules coming out of the engine bell will tend to bounce them back. These then strike other molecules coming out, pushing them back as well (but of course not all of them!). This effect provides a forward push on the rocket. Consequently, the total force acting on a rocket in flight in the atmosphere is:

$$F = u \times \frac{\mathrm{d}m}{\mathrm{d}t} + (P_4 - P_A) \times A_B$$

in which P_A is the atmospheric pressure, P_4 is the exhaust gas pressure and A_B the area of the exit hole of the engine bell. The first term in this equation is the *momentum thrust* that we are used to discussing. The second is called the *pressure thrust* and is due to the pressure difference between the front and rear of the rocket as discussed above.

At first glance this equation implies that the greatest force on the rocket would be developed if the exhaust pressure was much greater than atmospheric pressure. However, what is not obvious from the bare terms in this equation is that u (the exhaust speed) *also* depends on the exit pressure. A large exhaust gas pressure implies that there is not much flow to the gas. The molecules emerging from the bell have a large random motion (pressure) and therefore the overall velocity of the gas out of the bell (u) is comparatively small (the molecules are moving quickly in random directions, so overall they are not going very far!). Therefore, P_4 being large implies that u is small. Rockets are designed to consume propellant at a fast rate, hence the $\mathrm{d}m/\mathrm{d}t$ term is very large. The momentum thrust is a much larger contributor to the overall thrust than the pressure thrust. Increasing P_4 will, surprisingly, *reduce* the overall thrust as the smaller value of u this implies will have a much larger effect. Conversely, if the exit pressure is less than atmospheric then the pressure behind the rocket is less than the atmospheric pressure acting backwards on its nose. Furthermore, the exhaust gases have trouble exiting the engine bell in a smooth and linear flow. Overall less thrust is produced in this situation as well. The greatest thrust occurs when P_4 is equal to P_A and u is hence a maximum. For this reason engine bells are designed to bring the exiting exhaust gas up to a pressure that equals the external pressure, and so are most efficient at a given altitude.

NASA and Rocketdyne are currently experimenting with the *Aerospike* engine that works with great efficiency at any altitude without using an engine bell/nozzle system at all. We will discuss this design more in chapter 7.

Controlling thrust (throttling)

From the earliest stages of mission design, it was clear that one of the greatest technical challenges that NASA would face was the design of the lunar module descent engine. The demands of the powered descent and ability to fly the spacecraft about while looking for a safe landing site required an engine that could have its power output varied (throttleable). This had never been done before with a rocket engine.

During the development of the LM, NASA instructed the main contractor (Grumman) to sub-contract to two different companies for the preliminary design of the descent engine. Rocketdyne's engineers, who were responsible for the F-1 and J-2 engines already being used for the Saturn V, suggested that a throttleable engine could be produced by introducing some inert gas (such as helium) into the propellant, which would alter the mixture without changing the flow rate. This was a new approach to engine design, but would not require some way of adapting pumps to deal with a variable pressure. The rival design was put forward by Space Technology Laboratories Inc. (STL). Their idea was to use a pressure-fed hypergolic system. This design would not use pumps to deliver the propellant and so variable-flow-rate valves could be used to provide throttle adjustment. The design also called for an injector system that had a variable area (this is a similar idea to that used in some shower-head designs). Interestingly, Grumman eventually recommended that the Rocketdyne design be approved, but was overruled by NASA[10].

3.6 Staging

One of the often-quoted virtues of an active space programme is the pressure that is placed on the development of new technologies and materials. Materials that will survive the heat of re-entering the Earth's atmosphere have to be designed, as well as those from which rockets can be constructed with the smallest possible weight. For the Saturn V the Alcoa and Reynolds Corporations between them developed an aluminium alloy that was extremely strong and could be used in sheets at most 0.64 cm thick to make the external skin of the rocket. With such a small thickness of material involved, careful design was needed to ensure that the rocket could support the tremendous weight of the propellant. A fully fuelled Saturn V was some 20 cm shorter than when it was empty, simply due to the weight in the propellant tanks compressing the structure.

Once the choice of propellant has been made, then the specific impulse is fixed and high mass ratios become the goal of the design. Even then, as we have noted, in practice it is difficult to achieve a ratio between the masses of the propellant and the rocket that is much higher than 10:1. To make matters worse, we simply cannot stuff a rocket full of greater and greater amounts of propellant in the hope of achieving a higher final velocity. Physically the reason is simple. The more propellant in the rocket the heavier it is, and the harder the engines have to work in order to lift it off the ground.

Figure 3.8 makes the point graphically. On this graph N is the ratio between the total mass of the rocket (i.e. casing and propellant, and payload) and the mass of the payload that we are trying to launch. Along the curve the mass of the fully fuelled casing (i.e. not counting the payload) has been fixed at 10 times that of the empty casing. The exhaust speed is constant at 3 km s^{-1}. The curve shows what happens if we launch the same payload using bigger and bigger rockets. Even with impracticably large rockets 700 times the mass of the payload, the final velocity has not increased by much over that achieved with a rocket 50 times the mass of the payload.

The fully fuelled Saturn V sitting on the launch pad had a total mass some 56 times that of the Apollo spacecraft.

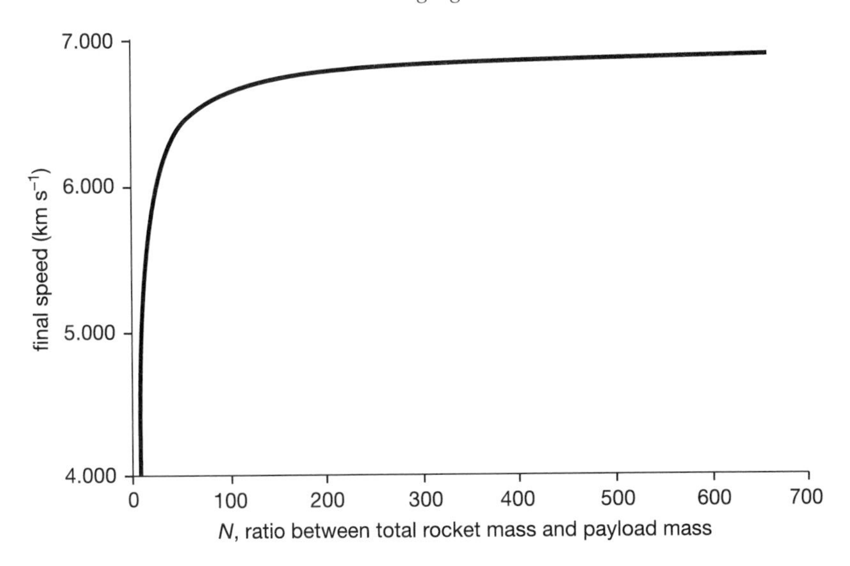

Figure 3.8 Final speed for a single-stage rocket, exhaust speed 3.0 km s^{-1}.

One can see quite clearly on this graph that no matter how heavy the rocket becomes, it is never going to get to a 7 km s^{-1} final speed. Single-stage rockets are fundamentally limited by the mass ratios that can be achieved.

The answer is to construct a rocket from multiple stages, each of which has its own system of fuel tanks and motors. At first glance this appears to be defeating the object as we are adding the mass of engines for each stage and so reducing the mass ratio in the rocket. However, this is more than compensated for by the ability to throw away the empty stage with its engines and, vitally, the empty fuel tanks that are no longer required. Then, as the second stage builds its ΔV on top of that already achieved by the first stage, an overall improvement in performance can be gained. It is a little like throwing a ball out of a moving train. The ball ends up with a considerable speed as it already has the speed at which the train is moving. If each stage has a mass ratio of 10:1 the performance is better than that of a single stage with the same ratio.

The first stage of a multi-stage rocket has the most powerful engines and carries the largest mass of propellant by far (see the maths box on multiple-stage rockets). First stages are generally designed to carry the rocket up into the higher atmosphere where the density of air is less and so the air resistance is considerably reduced. The second stage then fires and carries the payload almost into orbit. In the case of the Saturn V a burn from the third stage was then required to accelerate the Apollo up to the required speed for Earth orbit.

Single-stage rockets

One can see the limitations of a single-stage rocket by fiddling with the ΔV equation. The easiest way of doing this is to work in terms of multiples of the payload mass, as I have in figure 3.8. For the sake of simplicity I will use the equation that does **not** include the effect of gravity, but the conclusion will be the same. I will assume that the payload carries no propellant (or that the propellant it contains is not used as part of the launch sequence—as was the case with Apollo) and that it has a mass m. The casing mass (i.e. the empty rocket without payload or propellant but including propellant tanks and engines) is M_C, and the propellant mass is M_P. The total mass of the whole rocket, including the payload, engines, propellant and casing, is Nm where N is the multiple of the payload mass (as used in figure 3.8). Finally, I will set R to be the ratio between the mass of the empty casing and the mass of the fuelled casing. It is assumed that $R = 0.1$ for most practical rockets.

Summarizing

Mass of payload $= m$.
Mass of whole rocket at launch $= Nm = m + M_C + M_P$.
Mass of fuelled casing $= Nm - m = m(N-1)$.
Mass of casing $= R \times$ mass of fuelled casing $= Rm(N-1)$.
Mass of rocket after propellant is used up $= M_R = M_C + m = Rm(N-1) + m$.

Putting all this into the delta-V equation gives:

$$\begin{aligned}\Delta V &= I_{sp} \times g \times \ln\left(1 + \frac{M_P}{M_R}\right) = I_{sp} \times g \times \ln\left(\frac{M_P + M_R}{M_R}\right) \\ &= I_{sp} \times g \times \ln\left(\frac{Nm}{Rm(N-1)+m}\right).\end{aligned}$$

When the dust settles, this boils down to a remarkably simple expression:

$$\begin{aligned}\Delta V &= I_{sp} \times g \times \ln\left(\frac{N}{R(N-1)+1}\right) \\ &= I_{sp} \times g \times \ln\left(\frac{N}{RN}\right) = I_{sp} \times g \times \ln\left(\frac{1}{R}\right).\end{aligned}$$

In order to get to this I have used the fact that once N becomes as large as 1000 or more, then the difference between N and $N - 1$ may as well be ignored. Similarly, RN becomes at least 100 and $RN + 1$ is practically the same as RN. The final expression shows that the ΔV that can achieved with an enormous rocket does not depend on the mass of the rocket and is limited by the mass ratio. In this instance, with an exhaust velocity of 3.0 km s^{-1} and $R = 0.1$, the maximum ΔV is 6.9 km s^{-1}. Interestingly, this is less than the speed required to escape from Earth's gravity.

Multiple-stage rockets

To see the advantage of multiple-stage rockets in more quantitative terms, consider a two-stage rocket in which each mass is quoted as a multiple of the payload.

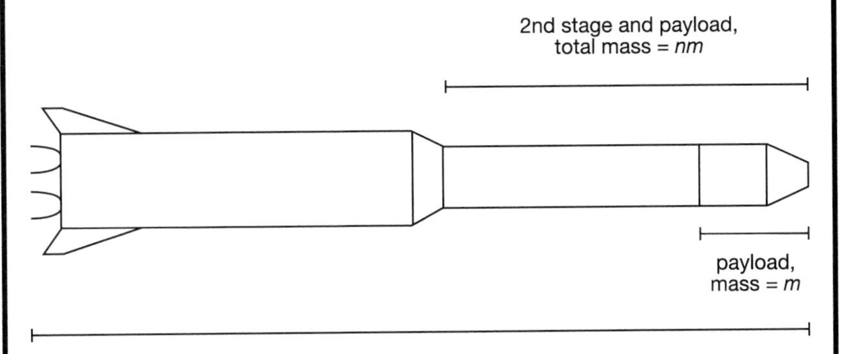

Once again R is the ratio of the mass of the empty casing to the mass of the fuelled casing.

Mass of 1st stage with propellant $= Nm - nm$.
Mass of 1st stage without propellant $= R(Nm - nm)$.

$\therefore$ after the 1st stage has burnt out, the mass of unseparated rocket $= nm + R(Nm - nm) = nm(1 - R) + RNM$.

$$\therefore \Delta V_1 = I_{sp} \times g \times \ln\left(1 + \frac{M_P}{M_R}\right) = u \times \ln\left(\frac{M_P + M_R}{M_R}\right)$$

$$= u \times \ln\left(\frac{Nm}{nm(1 - R) + RNm}\right) = u \times \ln\left(\frac{N}{n(1 - R) + RN}\right)$$

which does not depend on the payload mass at all.

A similar calculation produces the ΔV achieved by the second-stage burn:

$$\Delta V_2 = u \times \ln\left(\frac{n}{R(n - 1) + 1}\right).$$

Now, the total ΔV after the two stages have burnt out will be the sum of those achieved by each stage in turn:

$$\Delta V = \Delta V_1 + \Delta V_2$$

which depends on the relative sizes of N and n.

Multiple-stage rockets (continued)

The graph below shows how the total ΔV for a two-stage rocket varies with the mass of the second stage. The total mass of the rocket has been fixed as Nm with N equal to 50. Consequently, the mass of the second stage, nm, must be such that n is less than 50.

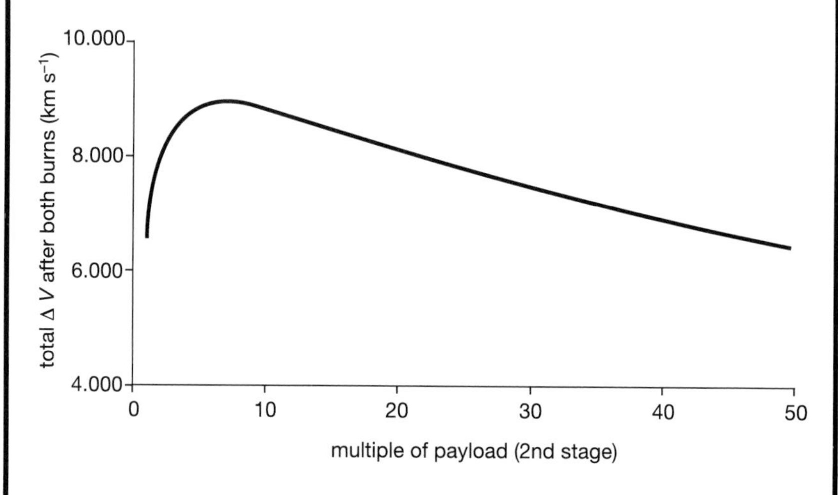

The graph clearly shows a peak at a low mass value of n. In other words, the largest overall speed is achieved if most of the mass is in the first stage. This is partly why RP-1 was chosen as the fuel for the Saturn V first stage—it is much denser than LH_2 and so a high fuel mass could be achieved in the first stage without a large volume tank required to store it.

A detailed calculation, using calculus, shows that the optimum is achieved if n is equal to the square root of N.

For a rocket with an overall mass equal to 50 times the payload mass, the greatest ΔV comes about if the first stage has a mass of $\sqrt{50}$, or 7.07, which is where the peak is on the graph.

Interestingly, fixing $n = N$ also has the effect of making the ΔV achieved by each stage in turn the same.

However, we must remember that this is not taking into account the role of gravity and the enormous air resistance that acts on the first stage in particular (as it generally flies through the thicker atmosphere).

In the specific case of the Saturn V, the three stages performed as detailed in the table below. Comparing the mass of propellant with the casing mass for each stage drives home the point about the large mass ratios that were used.

Stage	Burn time (s)	Propellant mass (tonnes)	Casing mass (tonnes)	Velocity at burnout (km h^{-1})	Altitude at burnout (km)
1	170	2147	136	9 654	61.1
2	395	444	40	24 617	184.2
3	165 (into orbit)	107	13	24 617	185

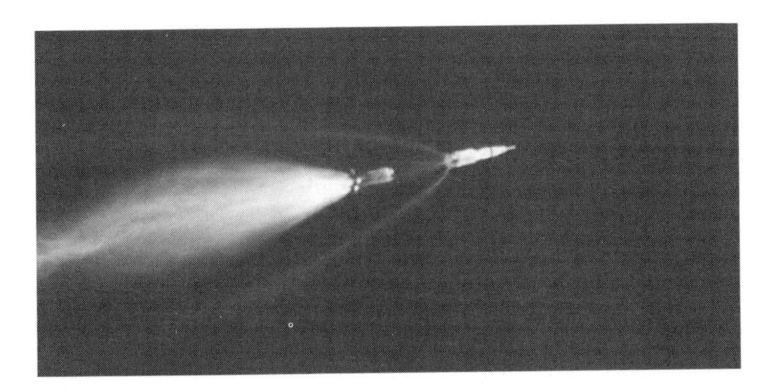

Figure 3.9 Staging of a Saturn V rocket—the exhaust from the second stage motors cannot be seen as it comprises water vapour!

3.7 A typical Saturn V launch

Propellant loading of the Apollo spacecraft (i.e. the CM and LM) was performed on the pad, prior to launch day. Aerozine 50 was used as the fuel, with nitrogen tetroxide as the oxidizer.

The process of readying the Saturn V itself began the day before launch by loading the first stage with the RP-1 fuel it required. This was pumped in through a 15 cm duct at the bottom of the 768 000 litre fuel tank. The fill rate was 760 litres per minute until the tank was 10% full, after which the rate was jacked up by a factor of 10 to 7600 litres per minute. Helium was used over the fuel to maintain a constant pressure in the tank both at launch and during flight. Without this it would be difficult to maintain a constant flow of fuel to the engines. The helium used for this was stored in four 0.88 m^3 bottles inside the LOX tank.

The cryogenic propellants (LOX for all three stages and LH_2 for the second and third stages) were loaded from about 7 hours prior to launch. If the tanks were at normal temperature most of the liquid would boil into a useless gas as soon as it touched the metal of the tanks. Consequently all the cryogenic tanks and pipes were pre-cooled by pumping cold gas through the supply lines. The pre-cooling of one tank could take place while another was being filled.

LOX was the first cryogenic propellant to be loaded. First the second stage was filled to 40% of capacity, followed by the third stage being completely filled. Next the second stage was brought up to capacity, followed by the first stage being fully filled. Splitting the filling in this manner allowed various leak checks to be performed while other tanks were being filled. It also allowed the rocket to settle under the stresses produced by the mass of propellant.

The next step was adding the LH_2, starting by fully filling the second stage and then the third stage.

In total it took four and a half hours to fill the rocket with its cryogenic propellants.

Filling the LOX tanks was a delicate process. To avoid damage from LOX splashing about inside, the first-stage tank was filled at a comparatively slow rate of 5600 litres per minute. Even with the pre-cooling of the tanks, much of the LOX arriving initially boiled away. Once the tank was about 6.5% full it had cooled sufficiently to prevent any further liquid boiling as it arrived. At this point a visual inspection took place to ensure that there were no leaks. If all was well, filling continued at 38 000 litres per minute until the tank was 95% full, after which the fill rate was reduced back to the earlier slow rate to top up the final level—1.25 million litres. With the rocket sitting on the pad, LOX would continually boil inside the tank which had to be allowed to escape into the atmosphere. The tank remained connected to the LOX supply until 160 seconds before launch so that the contents could be continually topped up.

The second-stage tanks were filled at 1900 litres per minute for LOX and 3700 litres per minute for LH_2 up to the 5% level. After this filling proceeded at 19 000 litres per minute and 38 000 litres per minute respectively. As with the first stage, the tanks were continually kept topped up—LOX until 60 seconds before launch and LH_2 until 70 seconds before launch.

At approximately 160 minutes before launch the astronauts entered the spacecraft. 45 seconds before engine ignition, helium was pumped through the LOX filling lines to pressurize the first-stage tank sufficiently to start the engines and build up thrust. After lift off, when the supply of helium was no longer available, gaseous oxygen was used to top up the volume of gas in the LOX tank. Without this, as the volume of liquid in the tank dropped, the pressure of gas over the liquid would reduce and the flow of LOX would drop.

Lift off commenced with a start signal to the five F-1 engines of the first stage. The engines were arranged across the base of the stage in the same pattern as the dots on a number 5 domino. The centre engine started first and then the opposite pairs fired up at 300 millisecond intervals. The engines were started by opening valves to allow LOX into the thrust chamber of each engine. A hypergolic solution was then injected into the chamber, where it reacted with the LOX. The main fuel valves opened and the RP-1 entered the thrust chambers to ignite and sustain the reaction started by the hypergolic solution. At this point engine thrust built up rapidly. However, the rocket was not allowed to leave the launch pad until full thrust was developed by each engine. Large clamps then released the stage, allowing it to rise, pulling long tapered pins through holes—this provided some restraining force so that the actual lift off was a comparatively gentle affair. Having said that, this 'soft release' mechanism only held the rocket for 500 milliseconds!

The Saturn V rose vertically off the launch pad for about 131 m, until it was well clear of the support arms from the gantry. It then began to pitch over and roll into the correct attitude for the rest of the flight. The four outboard engines were mounted to the stage on large universal joints that allowed the whole engine to pivot (gimbal) up to 6° in any direction. The flight path was constantly monitored and corrections were applied by gimballing the engines.

Some 69 seconds after lift off, air resistance acting on the rocket reached a peak as the speed built up to the speed of sound. At this moment the drag acting on the rocket was up to 2 million newtons. Having smashed through the sound barrier the astronauts reported that the ride smoothed out and became much quieter (obvious really when you think about it!). Most of the LOX and RP-1 was consumed by 135.5 seconds into the flight. At this moment the centre engine was shut down, while the remaining four continued to burn until all the remaining propellant was used up, at which point the computer controlling the flight cut off the remaining engines. Stage separation was achieved by a small explosive charge. Directly after stage separation, eight small rockets mounted on the first stage fired to slow its motion down and to allow the second stage to move ahead. The first stage's momentum carried it up to an altitude of 111 km before it fell back to Earth and crashed into the Atlantic ocean about 560 km from the launch point.

Meanwhile, eight solid rocket ullage motors on the second stage fired to settle the propellant so that the pumps could work effectively. Once the engines of the second stage had reached 90% of their full thrust, the inter-stage ring separated. This ring covered the engine bells of the second stage and formed a connection with the first stage. Jettisoning the ring was an incredibly precise manoeuvre. The 5 m tall ring had to slide past the engine bells with a clearance of 0.9 m.

About 30 seconds after ignition of the second stage the escape system was jettisoned from the top of the command module.

The second stage then burned for a total of 395 seconds, boosting the Apollo spacecraft nearly into orbit. When its propellant had been exhausted it separated in a similar manner to the first stage. Four solid-fuel rockets mounted around the inter-stage assembly assured a clean separation of the third stage from the second stage by slowing down the spent booster. The third stage took over after two solid rockets had fired to settle the propellant in its tanks. The stage burned for 165 seconds, depositing the spacecraft into orbit.

The next stage of the mission was to orbit the Earth several times while the systems were thoroughly checked out. This was especially important during the Apollo 12 mission—the Saturn V booster was struck by lightning during the launch, which lit up virtually all the alarm lights to do with the electrical systems. Although the booster and its trajectory were unaffected by the strike, the surge tripped out the service module power cells. Reconnecting them cleared the alarms. Nevertheless, the astronauts orbited the Earth for over two hours while the systems were thoroughly checked. Eventually they received the OK for the trans-lunar injection burn.

What mission control did not tell them, however, was that they were worried that the small explosive charges that were designed to deploy the parachutes for the splashdown of the command module had been damaged. In the end the mission controllers took the very pragmatic view that they may as well continue the mission to the Moon. After all, it would make no difference if they brought the men back straight away and the parachutes did not open, or if they let them fly to the Moon first. Fortunately their worries were unfounded and the astronauts returned safely to Earth at the end of the mission.

3.8 Future developments in rocketry

In the future it is very likely that the chemical rocket will be abandoned for all applications except launches from the Earth's surface and short range hauls (as far as the Moon). This is largely because of the limitations on the specific impulse obtainable from chemical means. The best chemical rockets have a specific impulse of 450 seconds, so in order to achieve the ΔV needed to explore the solar system, large mass ratios would have to be employed. This is unfortunate as solar system exploration will require sustaining astronauts for

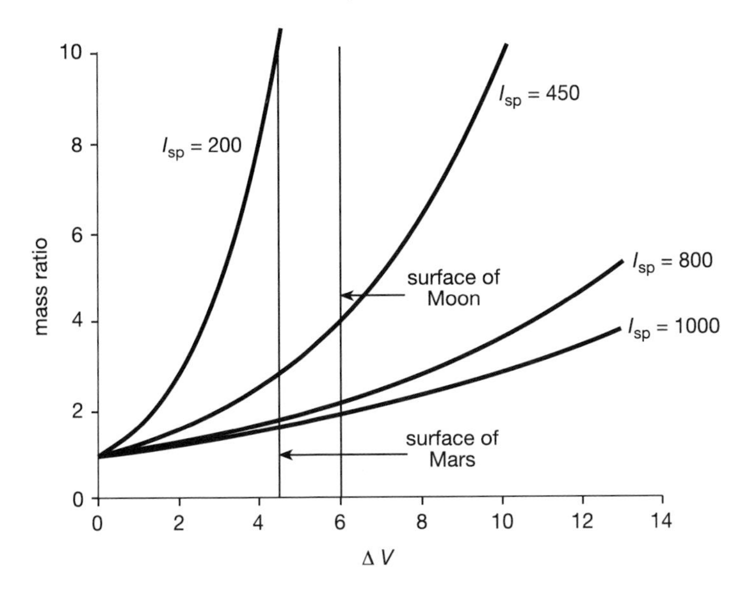

Figure 3.10 How mass ratio required for a given ΔV varies with specific impulse (the reason why the ΔV is less to get to Mars than to the Moon is explained in chapter 4).

long periods of time (measured in months, not days as for Apollo), and so the actual payloads will be very much heavier (consumables such as oxygen, water and food make up a substantial fraction of the mass). If the payload is heavier, then to achieve a given mass ratio the total spacecraft mass has to be very much heavier as well.

The goal of engine developers is to increase the specific impulse as much as possible, allowing the mass ratio to come down so that more of the mass of a spacecraft can be devoted to the actual payload. This is illustrated in figure 3.10 which shows how the mass ratio required to achieve a given ΔV varies with different specific impulses. Remember that in practical terms a mass ratio much greater than 10:1 cannot be achieved. There may be some hope for increasing this slightly as lower density alloys and composite materials for use in constructing rockets become available (see chapter 7). However, this is not going to push mass ratios up as high as 100:1.

If the specific impulse could be increased towards 1000 seconds this would have a dramatic effect on the mass ratio required.

Increasing the specific impulse implies producing a greater exhaust velocity. In the types of engine considered below, this is done by using the lightest mass propellant possible—hydrogen. Consequently, specific impulses of 800–1000 seconds are possible.

The flip side of this, however, is that these designs do not consume propellant at a very fast rate—the thrust is very small. Remember that ΔV depends on exhaust velocity and the mass ratio—not the rate at which propellant is used. An engine can use up propellant at a slow rate (low thrust) yet still build up considerable speeds provided that the thrust is maintained for long periods of time. These engines would be hopeless at lifting loads off the ground, but in space they can build up enormous speeds by gently thrusting away for most of the journey[11].

In chapter 9 we will take a look into the more distant future to see what forms of propulsion might be used for deep space and interstellar travel.

Nuclear engines

A nuclear reactor is an effective way of generating heat. Nuclear fission reactors work by splitting the nuclei of very heavy elements into pieces comprised of lighter elements, a process that releases energy. One common fuel used is uranium. Some varieties of uranium nuclei will split when a slow-moving neutron collides with them:

$$\text{neutron} + \text{uranium nucleus} \xrightarrow{\text{splits into}} X + Y + \text{some number of neutrons}$$

where X and Y are the nuclei of some lighter elements (it is impossible to say what X and Y will be in any individual case: the splitting process is random). Also produced are two or three neutrons, some energy in the form of gamma rays and some exotic particles called neutrinos.

The neutrons produced by the reaction can go on to collide with other uranium nuclei themselves, and cause further reactions, which will in turn produce more neutrons, etc. This continuing process is called a *chain reaction*. Once a chain reaction is started, by introducing some neutrons from outside, then it becomes self-sustaining (provided the situation has been engineered correctly—the neutrons produced have to have a good chance of colliding with other uranium nuclei, for example). The reaction is regulated by the use of *control rods*—cylinders of a neutron-absorbing material that can be lowered into the region where the uranium is reacting. The further in the control rods are pushed, the greater the number of neutrons that are absorbed and the slower the reaction rate, and vice versa.

Most of the energy produced in the reaction is the kinetic energy of the products X and Y. These nuclei are produced moving at considerable speeds. In a power station reactor the uranium fuel is normally stacked in disks and contained in fuel rods. Consequently X and Y cannot go very far without colliding with other nuclei inside the fuel rod. After several such collisions their kinetic energy has been spread throughout the rod, raising the temperature. After a short while the temperature of the reactor as a whole becomes very high. Coolant circulating through the reactor extracts thermal energy and carries it away to where it can be safely used. In the case of an electricity generating station the hot coolant is used to boil water into steam which drives the turbines that generate the electricity.

A somewhat similar process could be used to produce thrust for a spacecraft. A compact nuclear reactor, such as those found on nuclear submarines (which have a slightly different operating principle to that described above), would be used as a heat source. The thermal energy extracted from the reactor would boil a propellant liquid. The escaping gas could then be used to provide thrust.

Remember that the principle of rocket propulsion simply requires high-speed molecules to push against the rocket. In a chemical rocket these molecules are the products of a chemical reaction. In a nuclear engine they are simply the propellant gas boiled off stored liquid. In outline the idea is rather like using the steam produced from a kettle to push the kettle across the room. On Earth this would not work, but in space a kettle would gently push itself along with its escaping steam.

A suitable propellant for a nuclear engine would be liquid hydrogen, comparatively small volumes of which can be stored in liquid form. Not only that, but hydrogen is the lightest element, so for a given temperature its molecules will have the highest speeds. The low temperatures required to keep it in liquid form are much more easily maintained in space than on the ground. Such a rocket would never produce enough thrust to lift a spacecraft off the ground. However, it would be excellent propulsion for trips between planets as the gentle thrust could be maintained for very long periods (weeks!) and so considerable speeds could finally be built up.

In the 1960s development work on nuclear engines was carried out by NASA. The Nuclear Engines for Rocket Vehicle Applications (NERVA) programme constructed and tested several such engines with thrusts between 44 000 N and 1 111 000 N. These engines delivered specific impulses of the order of 800 seconds—vastly superior to chemical rockets. At that stage the engines were being developed with a manned Mars mission in mind. Unfortunately, when the Nixon administration cut the NASA budget, NERVA was one of the casualties. However, the principle of nuclear engines was successfully tested and they remain good candidates for long-term development in interplanetary spacecraft.

Despite their evident virtues, nuclear engines do have two drawbacks.

Firstly, it is difficult to arrange for the heat exchange from the reactor coolant to the liquid hydrogen to be done efficiently. A great deal of energy is wasted in this process. Secondly, as I noted above, nuclear engines could not be used to lift a spacecraft off the ground. Therefore at some point a nuclear reactor, or the components for constructing one, would have to be launched by standard chemical rocket.

So far environmental lobbies have been extremely reluctant to allow radioactive materials to be flown in a chemical rocket. The prospects for scattering such material across a wide area if the rocket should explode are very serious[12]. Some small reactors have flown as fuel cells on deep-space probes, but nothing near the size that would be required for a nuclear engine.

Solar thermal

The solar thermal engine uses the same basic idea as the nuclear engine—boiling a propellant liquid to provide thrust. However, the heat source in this case is focused light from the Sun. This is evidently much more practical in space where there are no clouds to block the view of the Sun. This engine also has the advantage of not needing radioactive materials to be launched into space. However, in order to generate the temperatures required, quite concentrated sunlight is needed which makes it difficult to use in the outer solar system. Also projections indicate that the engine would only be able to develop about 400 N of thrust. Nevertheless, some development is being carried out at the moment as a 'greener' alternative to nuclear engines.

Ion motors

Like the nuclear engine, the ion rocket would be useless at lifting objects off the ground. However, it can also be used to provide gentle thrust for considerable lengths of time. The idea behind the ion motor is to make use of the electrical properties of ionized atoms. Normally an atom is electrically neutral as it has an equal number of negatively charged electrons to the positively charged protons in its nucleus. Under some circumstances an atom can have one of its electrons removed to become a positive ion (now it has more protons than electrons so that it is net positive). Negative ions can also be formed when an atom has an extra electron bolted on.

Ions can be steered using the electrical forces from other charges, or the magnetic forces from electromagnets. The idea behind an ion engine is to accelerate ions to very high speeds and fire them out of the back of a spacecraft. The principle of accelerating charged particles in such a manner is already well established on Earth.

The most appropriate way of accelerating charged particles for use in a propulsion system is to use a linear accelerator. This consists of a series of metal 'cans' aligned with a central hole running right through the set. If the particle to be accelerated is positively charged (say a proton), then the first can is set up with a negative charge, attracting the proton towards it. As the proton coasts through the can it feels no pull from the charge on the metal surface (it is a general property of hollow, charged metal objects that they exert no forces on charged objects within them). The trick now is to switch the charge on the can over to being positive while the proton is passing through. Then, as it emerges from the can at the other side it is repelled away towards the next, negatively charged, can. By continually switching positive for negative in sequence down the line, a stream of protons can be accelerated. Linear accelerators of this type have been in use for many years to accelerate particles to very high energies for use in physics experiments. The largest is 30 km long, at Stanford (the Stanford Linear Accelerator Centre, or SLAC for short).

On Earth it is easy to ignore the role that Newton's third law plays in such an accelerator. As the proton is pulled and pushed from can to can, it exerts equal and opposite forces on the cans themselves. Of course, the equipment of the accelerator is firmly attached to the ground and so the change of the Earth's momentum that the proton's forces achieve is far too small to be measured. However, in a freely floating spacecraft the push of the protons back on the accelerator, and hence the spacecraft itself, is sufficient to provide a useable thrust.

The ion engine has considerable potential as it is capable of generating specific impulses of thousands of seconds. However, in order to do that a great deal of electrical power is required. In order to push a 120 tonne spacecraft with a force of 280 N at a specific impulse of 5000 seconds, about 5 megawatts of electrical power would be required. To get a feel for the size of this, remember that most power stations are 50 megawatts or more and that the projected power requirement for the international space station is about 7 kilowatts. The only way of generating such amounts of power is by using a nuclear reactor, so the ion engine suffers from the same environmental problems as the nuclear engine. Kilowatt-size ion drives do exist and some satellites use ion engines for manoeuvring purposes.

NASA's Jet Propulsion Laboratory (JPL) has worked out a design for a probe capable of reaching nearby stars based on an ion drive. The *Thousand Astronomical Unit*[13] probe (TAU) would use a nuclear reactor to generate

150 kW of electrical power and eject a stream of xenon ions at about 70 km s^{-1}. The probe would be capable of accelerating for ten years, by which time it would have reached a speed of 95 km s^{-1}.

On 24 October 1998 NASA launched DS1 (Deep Space 1), the first of twelve planned *New Millennium Program* (NMP) missions and the first space vehicle to use an ion drive as the primary means of propulsion. The purpose of the NMP is to test out new technologies in high-risk scientific missions. In other words, the missions carry a significant chance of failing due to technological problems, but if they work they will have helped development of the technology and returned some scientific data into the bargain.

It is intended that DS1 will visit asteroid 1992 KD during its mission and perhaps also comets Wilson–Harrington and Borrelly. At each encounter it will perform tests of its advanced instrumentation. The best way of testing equipment that might be used in even more demanding missions in the future is to try to use it to do real science. It also ensures that rare opportunities to encounter a variety of fascinating solar system bodies during a short mission are being fully exploited.

One of the later NMP flights, DS4, will carry four ion engines and will be targeted to explore a comet in detail.

The ion engine on board DS1 was manufactured by NSTAR (a cooperative project between the NASA JPL and NASA Lewis centres) and uses a beam of xenon ions accelerated through 1280 V. At peak power the engine provides 92 mN of thrust and uses 2.3 kW of electrical power, and has a specific impulse of 3100 s. The electrical power is provided by solar panels supplemented by storage batteries. Due to the limitations of the solar panels it is not expected that the engine will run at full power for much of the mission. In December it was found that the engine could run happily at 85% of full thrust, although this will decline as it gets farther from the Sun. At launch the craft was carrying 82 kg of xenon. In principle the ion engine should be able to deliver ten times the ΔV that a chemical rocket could with the same mass of propellant.

On 8 December the ion engine was turned off after two weeks of continuous operation—which was more than twice as long as it had been planned to run the engine without interruption.

Another intriguing technological aspect of DS1 is its autonomous guidance system. One aspect of this is on-board navigation. The computer will be able to tell where the craft is by taking photographs of asteroids and comparing their positions against the background stars. It has the locations of 250 asteroids and 250 000 stars stored in its memory. There is also software that gives the computer a much greater degree of control over the spacecraft's trajectory than has been possible before. With this new software the ground controllers will give the system more generalized tasks to accomplish and will rely on the system to do it in the best way that it can. With this in mind, the planners are hoping that it will be able to guide itself to within 10 miles of the target asteroid.

Notes

[1] When the first V2 crashed into London, Von Braun remarked to one of his colleagues 'the rocket worked perfectly except for landing on the wrong planet'.

[2] In order to prevent ice forming on the sides of space shuttle external tanks they are sprayed with an insulating foam. Lumps of ice falling from a tank during lift off, as they did in Apollo, would have a significant chance of damaging the delicate heat-protecting tiles on the Orbiter.

[3] During the war England argued that the Germans were unable to build a rocket due to this weight problem. They did not accept at the time that liquid-fuelled rockets were possible.

[4] By the rocket's casing I mean the structure, engines, propellant tanks, etc, i.e. the mass of the rocket with no propellant on board. The payload is not counted either.

[5] This is known as the Venturi effect. Old-fashioned car fuel systems used carburettors in which fuel was passed through a narrow throat (Venturi) reducing the side pressure it exerted. This drew air in from the side starting the fuel–air mixing process. Similarly, when air flows over the top of an aeroplane wing the molecules travel parallel to the surface so the pressure above the wing is reduced. Lift is generated by the greater pressure below the wing pushing it upwards. The effect is the reason why people are warned not to stand too close to a platform when a train is passing. The train drags air along with it, so molecules are travelling parallel to the tracks. Consequently the sideways pressure near the train is reduced and people can be pushed forwards as the train passes due to the greater pressure behind them.

[6] This always sounds terribly painful to me.

[7] Sound waves are variations in pressure moving through a gas. If the molecules are moving more quickly than pressure variations can move, then the gas will no longer be sensitive to such variations.

[8] Rocket engineers speak loosely of thermal energy being converted into kinetic energy in the engine bell. Of course, the thermal energy in the gas is a form of kinetic energy—the kinetic energy of the random motion of the gas molecules. What they are implying is that the bell is turning the random motion of the exhaust molecules into coherent motion along the line of flight. The pressure of the gas is also due to the random motion of molecules. When the engine bell is correctly 'tuned' the gas pressure will be equal to

the external atmospheric pressure. Some people may be wondering how the exhaust gets out of the bell! The point is that the molecules are coming out in a straight line (coherent motion), which is not pressure in the true sense. There is a subtle difference between the random motion of molecules and their coherent motion. In the atmosphere, the pressure is due to random motion, but wind is due to a degree of coherent motion.

[9] In this section I am not including the effect of air resistance (drag), which is another effect due to the flow of air over the rocket in flight. This is a static effect due to pressure differences.

[10] Both companies were progressing well and development was equally well advanced for both designs. NASA's decision may have been partly because they felt that STL could commit more resources and manpower to the project, unlike Rocketdyne which was also involved in F-1 and J-2 development.

[11] In practice, the engine would be used to increase speed for half the journey and then turned round to reduce speed for the second half.

[12] This is why it is not a good idea to dispose of nuclear waste products by firing them in rockets towards the Sun.

[13] As distances are so vast in space, astronomers sometimes use different units of distance. One such unit is the Astronomical Unit, which is the mean distance between the Earth and the Sun: 1.496×10^{11} m.

Intermission 2

From Mercury to Gemini

Having successfully placed the first American satellite in orbit, Von Braun's team was cut loose with the brief of 'man in space soonest'. The Mercury space programme was born. NASA set about selecting seven men who would form the basis of a new astronaut corps and the engineers began modifications to the Redstone and Atlas missiles. Redstone was not powerful enough to accelerate a manned capsule up to orbital speed, but it was capable of lobbing a man on a ballistic flight out of the atmosphere. Valuable experience would be gained operating in space, manoeuvring a capsule and during re-entry. Meanwhile the much more powerful Atlas would be readied for the orbital shots.

Compared to the Gemini and Apollo craft that were to follow, the Mercury capsule was tiny. It was just under 3 m tall and 1.9 m wide at the base, bell shaped with a pressurized compartment that was only just big enough to force a man into.

Mercury was the sort of ship that you wore rather than rode in. In 1961 Cape Canaveral was a scene of frantic activity. Test flights of the Redstone and Atlas rockets carrying unmanned Mercury capsules were taking place with varying degrees of success. The Redstone was proving to be quite reliable, but the Atlas was another matter. The material of its hull was so thin in places that it was only the internal pressures that kept it from collapsing. With its warhead replaced by a Mercury capsule, the rocket had an alarming tendency to fail during flight.

The climate of competition with the Soviets was intense. Soviet ambassadors and representatives seemed never to lose an opportunity of commenting in public about the relative achievements of the two space programmes. [In fact, while visiting the set for the film *2001: a Space Odyssey* in May 1966 the Soviet Air Attaché commented that all the labels on the controls of the American space ship *Discovery*, featured in the film, should be in Russian.] NASA moved determinedly, but cautiously. Alan Shepard was selected as the first of the

Figure I2.1 The tiny Mercury capsule and its escape tower.

Mercury Seven astronauts to ride the Redstone into space. His turn, however, had to wait. After successful unmanned flights the next stage was to launch a chimp called Ham. While the astronauts trained in simulators the chimps trained to bang levers in space—rewarded by banana pellets and punished by jolts of electricity. Ham was selected as the best of his team. The highly qualified, supremely fit and ambitious Mercury Seven did not take kindly to riding second after a chimp, but the doctors felt that there were too many unknowns about space flight. With hindsight it is difficult to appreciate some of the fears—would the astronaut's eyes distort under zero *g* and prevent them from seeing properly? Despite a vigorous campaign on the part of Alan Shepard to get the chimp grounded, the doctors insisted.

On 31 January 1961 the astronauts gathered at the cape to watch Ham take his flight. The chimp performed flawlessly. The rocket did not. The Redstone burned its fuel too quickly and the escape tower system, sensing that something was wrong, fired to lift the capsule clear of the rocket. The acceleration was brutal. Inside Ham proceeded to hammer on every lever that he was trained to hit, but the electrical system had gone wrong so he was greeted by a shock instead of a reward no matter what he did. The capsule sailed nearly 200 km further away from the Cape than anticipated and crashed into the water. By the time that the recovery helicopters arrived Ham was strapped into a capsule half full of water and nearly drowned. One can imagine the scene at the official greeting laid on for Ham when he returned to the Cape and emerged attempting to bite anything that came near to him.

The fault in the Redstone was traced to an electrical relay and Shepard was ready to take the next flight. Von Braun stopped him. Heavy with the extra responsibility of a man's life he wanted another test flight. That test flew perfectly on 24 March 1961. Shepard could well have been on board. Yuri Gagarin made one complete orbit of the Earth on 12 April 1961 while Shepard's Redstone sat on the pad having final adjustments made. The propaganda coup was crushing. In the middle of the night, as Gagarin passed over America, reporters were looking for some comment from NASA. One official, woken by a reporter, replied 'if you want anything from us you jerk, the answer is we're all asleep'. The next morning's headlines read:

SOVIETS SEND MAN INTO SPACE;
SPOKESMAN SAYS U.S. ASLEEP.

Two days later President Kennedy pulled his advisors into his office—how could America catch up? He gave his Vice President Lyndon B Johnson the task of finding a way to beat the Soviets. In the meantime the American programme seemed to be falling apart. Another Atlas failed and a smaller rocket designed to help test the Mercury escape system spun and crashed. Top people in the White House urged the President to cancel the manned programme. It was too dangerous. The publicity had been bad so far; imagine what would happen if an American died on top of an American rocket. Johnson urged the President to continue. The failings had no relevance to the Redstone rocket—they needed to buy time to get the Atlas right. Shepard took his flight on 5 May 1961. It lasted 15 minutes 22 seconds. Twenty days later Kennedy announced to Congress that they would go to the Moon.

The Mercury programme continued. The second flight was another sub-orbital hop on a Redstone. It was a great success until shortly after splashdown, when the capsule hatch mysteriously blew open, filling the capsule with water. The astronaut, Gus Grissom (later to die tragically in the Apollo 1 fire), scrambled out while the craft sank.

The first Atlas flight carried John Glenn into orbit and an adventure that made him an American hero (see intermission 6). There followed a further three Mercury flights, culminating in Gordon Cooper's 22 orbits on 15 May 1963. This was another flight that ended with electrical problems on board and the pilot having to fly the craft through re-entry manually. He splashed down just four miles from the recovery carrier. Despite this flight stretching the Mercury capsule to the limits, it was not clear that the Americans were catching up with the Soviets. On 14 June 1963 the Soviets put a man into orbit for a staggering

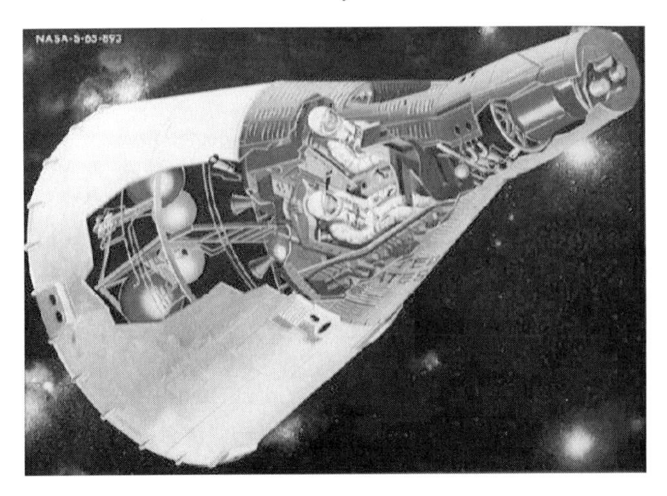

Figure I2.2 The Gemini capsule.

119 hours. Two days later they followed this up with the first woman in space. On the American side, things went rather quiet.

The decision to go to the Moon had kick-started a debate within NASA as to how best to achieve such a goal. The Apollo programme had Congressional funding and Von Braun's team were busy working on booster configurations, but the way in which the flight would be carried out was a matter for some debate. The matter was settled in 1962. Despite early reservations, NASA decided to go for a mission profile that required two spacecraft to find one another in orbit and dock. Not only that, but this crucial phase of the mission would take place in orbit round the Moon. Clearly NASA and the astronauts (now increased in number to 15) had to become experts in rendezvous and docking—quickly. This was the philosophy behind the Gemini programme.

Gemini was a much larger two-man capsule that was designed to be highly manoeuvrable in space. Mercury had shown that a capsule could be steered in its path, but Gemini was designed to be able to change orbits and dock with other craft. While America readied Gemini there was an opportunity for the Soviets to step in again.

In October 1964 the Soviets flew the first three-man mission. On 18 May 1965 Aleksey Leonov performed the first space walk—five days before the first Gemini flight, the date of which had been announced in February.

The Americans had not been totally quiet during this time. Unmanned probes were being sent to study the lunar surface to prepare for Apollo to follow.

Yet it looked as if nothing significant was happening. In the midst of this hiatus President Kennedy was assassinated. The programme had lost the man responsible for setting it in motion. Fortunately President Johnson was possibly an even more passionate supporter of the space programme.

The first manned Gemini flight took place on 23 March 1965 carrying Gus Grissom, who became the first man to fly in space twice, and John Young. Over the next year and a half a further nine Gemini flights pushed NASA to the front of the space race. They perfected the techniques of rendezvousing with robot target craft and with other manned Gemini capsules. Space walking was perfected, although not without some difficulty. In case of a docking failure it was vital that the Apollo crew could transfer from the lunar module to the command module by a space walk. Experience on Gemini showed that it was very hard to get a grip on the smooth sides of a spacecraft and that simple tasks that could be accomplished on Earth often took considerable time and effort in space.

Gemini was not without its drama. The flight of Gemini 8, carrying David Scott and Neil Armstrong (later to command Apollo 11 and become the first man to walk on the Moon), nearly ended in disaster when one of the manoeuvring thrusters jammed open, putting the craft into a spin so violent that the crew nearly blacked out. Only some very quick thinking on the part of Armstrong saved the mission. He used the re-entry thrusters to try to stabilize the ship while the jammed thruster ran out of propellant.

The last Gemini flight was a masterpiece. Shortly after launch the crew, Jim Lovell (later to command the ill-fated Apollo 13 and already a veteran of a previous Gemini flight) and Buzz Aldrin (later to land on the Moon in Apollo 11), discovered that their on-board computer system was not functioning properly. Fortunately Aldrin (the only member of the astronaut corps with a PhD) had worked out theoretically most of the manoeuvres needed to rendezvous with another craft in orbit. He was the acknowledged expert on such matters in NASA. Using a sextant and charts that he had produced in case of such an emergency, he guided the Gemini to the robot craft that they were due to find. Later his intense study of the problem of space walking, and the various improvised tools that he had carried, enabled him to carry out a series of complex tasks outside the craft. NASA became more confident that they could cope with the problems of space walks.

The scene was set for the Apollo programme to follow. However, NASA was about to face its first real tragedy. It would not be the loss of men on a mission, as had always been feared and which, as test pilots, the astronauts had always known would be a possibility. America would lose its first members of the astronaut corps in a fire on board the Apollo 1 capsule while it was sitting on a launch pad during a pre-flight test.

Chapter 4

Orbits and trajectories

The contents of this chapter cover gravity, centripetal force, the nature of an orbit and trajectories to the Moon. Weightlessness and training astronauts for zero g will also be discussed. Finally, we will come onto space stations—Skylab, Mir and the plans for a new major international space station.

4.1 Hollywood gets it right

Over the years writers of science fiction films have used many different literary devices to overcome one of the basic problems with trying to film people in space—the lack of gravity. Some shows barely mention the problem in passing, making some vague reference to artificial gravity (a nice idea, but with no conceivable prospect in physics). Others have used the idea of revolving sections of spacecraft to simulate gravity inside[1].

However, when you are trying to film a true story of what happened on board a real spacecraft, then you have to deal with the problem head on. This was the dilemma facing the producers of the film *Apollo 13*. Most events in the story take place within the Apollo command and lunar modules—they did not spin fast enough to produce any noticeable simulation of gravity inside[2].

Their solution was to fly the actors in an aeroplane and to use a skilled pilot to put the plane into *free fall*. This is a standard NASA training technique. An aeroplane is flown on a specifically curved path—a *parabola*. At the peak of the curve the plane dives towards the ground, still following the shape of the curve. Along this curve it is as if the plane were freely falling from a great height. As a result, anything that is not tied down inside the plane starts to float about. For the film the necessary sets were constructed inside the plane and filming took place during each run. In this manner a completely realistic simulation was made of the experiences of the astronauts on the way to the Moon. Yet, while the aeroplane was following this curve, the crew, actors and all the equipment on board were still under the influence of the Earth's gravity. This seeming paradox is only resolved by a full understanding of what is meant by 'free fall' or 'weightlessness' and how it is that objects can be placed in orbits that will circle planets indefinitely.

4.2 Falling again

In section 2.4 we discussed a surprising feature of the gravitational force—all objects, in the absence of friction or air resistance, fall at exactly the same rate. Free fall is the physicist's term for falling with only the force of gravity acting. If you were to jump off a diving board into a swimming pool, then on the way down you would experience a pretty close approximation to free fall. Air resistance would be very small as you would not be moving that quickly. However, were you to jump out of an aeroplane as part of a sky-diving team, then after a few moments you would no longer be in free fall. From this sort of height the speed at which you are falling would build up considerably (you have further, and so longer, to fall), and with it the air resistance. Sky divers spend most of their time before opening parachutes falling at a constant speed as the air resistance is equal to the pull of gravity on them—this is known as falling at *terminal velocity*[3].

In section 2.5 I explained how the couches on which the astronauts were lying provided the force required to accelerate them during lift off. I illustrated this idea by comparing the couch to a weighing machine comprising a large spring on which the astronaut is standing. The machine measures weight by recording the amount of compression in the spring. As the rocket takes off, the spring is compressed between the force due to the astronaut's weight acting down and the force of the engines transmitted upwards through the floor. Consequently the machine records a greater weight than if the rocket was stationary on the ground. We experience something very similar to the sensations felt by astronauts in this situation when we stand in a lift that is accelerating upwards rapidly.

Now consider what would happen to the machine if the rocket were falling towards the ground. For the moment let us forget about air resistance and assume that the whole set up is in free fall. This being the case, we can say that the astronaut, the spring and the floor are all falling towards the ground at exactly the same rate. Indeed, the astronaut is not really in contact with the top of the spring at all. He happens to be falling towards the ground just above the spring. Remember that in the stationary situation the spring is compressed as the astronaut is being pulled towards the ground by the force of gravity. The spring is held up by the rigid floor, so it is forced into compression by being trapped between two opposing forces. In free fall the astronaut is falling, but so is the spring, and the two are keeping exact pace with each other even though they have very different weights. The spring is not being compressed. The machine will read zero. In a sense the machine is reading falsely. The astronaut certainly still has weight, in the technical sense that he is still being pulled towards the Earth by the force of gravity. However, all the physical sensations that go along with our experience of weight will have disappeared.

Presumably, as you are reading this book you are sitting on a chair (or perhaps you are unfortunate enough to be standing in a London tube train being compressed on all sides by other people: if so you have my sympathy). What is it about your current situation that convinces you that you have weight? Physicists define weight as the force of gravity acting on an object. This force always acts at the centre of gravity of the object which, in the case of humans, is the middle of the body at round about the height of the navel. Now you certainly do not have some sensory organ that detects this force acting inside your body. So when we talk about weight in common terms we must be referring to something other than the very specific force that physicists have in mind.

At the moment I am sitting in a very comfortable chair bought for me by my family some years ago. My immediate experience of weight is the sensation of the chair pushing up on me. I also have some feeling that my lower legs are hanging down and pulling on my knees (the chair is tipped back slightly so my feet are not quite touching the floor). I certainly have tendons and muscles near my knees that can detect forces acting—they have to extend or contract in order to balance these forces. It is the sensation connected with this muscle activity that we associate with having weight.

A person standing on the ground is in a state of compression, just like the spring. They are being pulled towards the ground by gravity and being pushed upwards by the ground beneath their feet. Their joints are being compressed, as are their bones. In free fall, only the force of gravity is acting so the body is not in compression. The joints open out slightly—the sensation of weight disappears. This is what we mean when we say that an astronaut in orbit, or in free fall in the training aeroplane, is weightless. We do not mean that there is no force of gravity acting. Indeed no human has ever been in a situation where there was no force of gravity acting[4].

Inside a spacecraft in orbit round the Earth an astronaut could float weightless, hardly in contact with the floor. If she held an object in front of her and let go, then it would fall towards the ground—just as it would on the surface. The difference, however, would be that the person would also be falling towards the ground at the same rate, so the object would appear to float in front of them.

While travelling towards the Moon, the astronauts are still in free fall. Once the engines have been turned off the only force acting on them is gravity, and so they are still falling (for a while they are falling back to Earth, then once they get nearer to the Moon they start falling towards it). One of the ideas that people find hardest to understand is that an object can be falling back to Earth while still moving away from it. Just think about throwing a ball up in the air. All the time it is going up the Earth is pulling it back.

A freely falling aeroplane is not really a simulation of what is happening in orbit, it is a duplication of the conditions—except over a shorter time period.

When I am teaching about this in the classroom, I can always tell the moment when my students have grasped what is going on. Real understanding comes with the realization that a spacecraft cannot remain in orbit unless the force of gravity is acting. If, as they suspect from the expression 'weightless' which they know applies to astronauts in orbit, there was no force of gravity acting, then according to Newton's first law the spacecraft should be moving in a straight line with a constant speed!

A spacecraft in orbit (and all the people and objects within it) is in a continual state of free fall. Above the atmosphere there is no air resistance no matter what the speed. The whole thing is constantly falling towards the ground—but it never actually gets there[5].

4.3 Orbits

Newton came up with a nice illustration of how an object could continually fall towards the ground and yet never actually get there. His method of explaining the idea starts with simple gunnery, but we will consider golf instead.

At the end of the last Apollo 14 Moon walk, the Mission Commander Alan Shepard took his sample collector, connected the head of a six iron golf club to the end and, in front of a bemused TV audience, hit the first golf shot on the Moon (one-handed as his bulky space suit did not allow him to get a proper grip on the club). With the Moon's gravity being only 1/6th of that on Earth, and in the absence of air resistance, the ball travelled a considerable distance.

The path followed by a golf ball in the absence of air resistance, and without spin to complicate the issue, is referred to by mathematicians as a *parabola*. Any object that moves horizontally at a constant speed, and accelerates vertically at a constant rate, will follow a parabolic path.

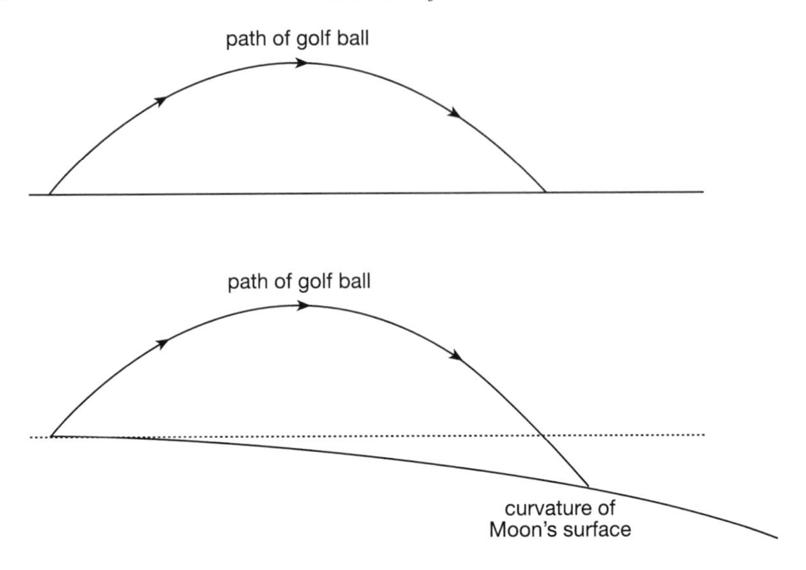

Figure 4.1 In order to calculate the true distance travelled by a golf ball on the Moon, the curvature of the surface must be taken into account.

Such a curve is shown in figure 4.1. The top picture is that of a golf ball in flight on the Moon. It is a simple matter for those familiar enough with the physics involved to calculate how far the ball would travel given the speed at which it left the club and the angle at which it started to fly. However, this would not give a true distance unless the curvature of the ground was also taken into account. On Earth such matters hardly bother golfers. Earth is a big planet, so the curve of its surface is quite small over regions the size of golf courses. However, when you are firing missiles from one part of the world to another, or at least planning to fire them, such things must be taken into account. As is clearly shown in the lower diagram, the ball will actually cover a slightly greater distance as the ground is not flat. In fact, if we do regard the planet as being curved, then strictly speaking the path of the ball is not a parabola. This is because the pull of gravity is always towards the centre of the Earth. Hence as the Earth curves so the direction of the gravitational force changes. For the path to be an exact parabola the direction of the force acting must not change.

In figure 4.2 three different 'parabolic' paths are shown. The difference between the three is that the ball has been hit harder and harder (but at the same angle to the ground). The faster the ball leaves the face of the club the less curved the path, and the further the ball will travel.

The next step is to change from a six iron to a putter and to imagine putting off the top of a mountain. If the ball is struck horizontally (i.e. parallel to the

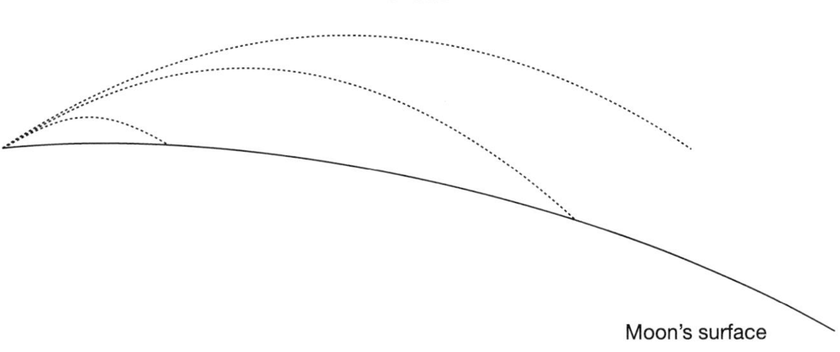

Figure 4.2 Striking the ball at the same angle, but with different speeds.

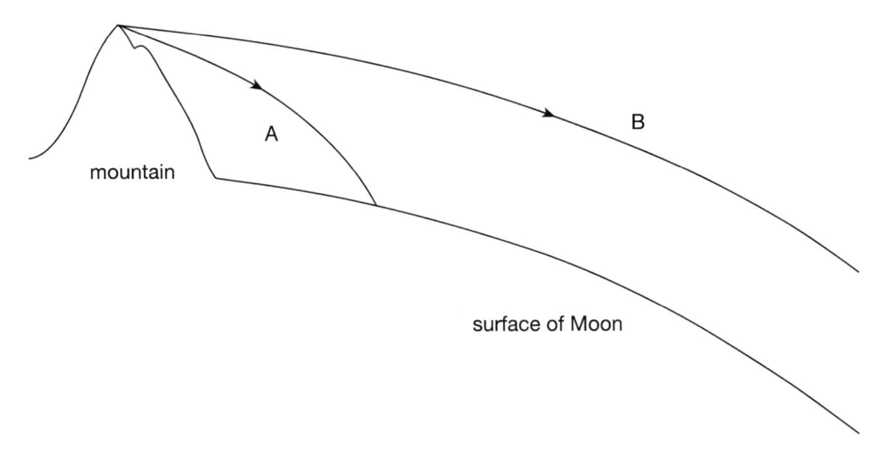

Figure 4.3 A ball putted off the side of a mountain will fall to the surface if hit too slowly (A), but with the correct speed will be sent into orbit (B).

ground at that place), then it will roll off the top of the mountain and fall in a path that corresponds to part of the equivalent path for a six iron shot—if the planet were flat it would be half of a parabola.

As with the six iron shot, the faster the ball moves the less steeply will its path be curved. In fact, with a ball moving sufficiently quickly **the curve of its path can match the curve of the planet's surface**.

When this happens, the ball will be continually falling towards the surface, which will in turn be curving away from it along a parallel path. So the ball continually falls to the surface without getting any closer. The ball is in orbit. Not only that, it is in free fall. If the same situation were to be reproduced in a spacecraft it would also be in free fall and all the occupants would be 'weightless'.

Now of course, in practice, putting a golf ball into orbit from the side of a mountain would be impractical even on the Moon. For one thing, at such a low altitude it would be very difficult to produce a course that would not have the ball hitting other mountains on the way. The point of this story is to illustrate what an orbit is. In practice, rockets launched from the surface follow a particular trajectory into orbit which ensures that by the time they get to the chosen height they are moving parallel to the ground and at the correct speed. With this done, gravity will continually swing the spacecraft around the planet in a constant orbit.

4.4 Centripetal forces

One of the things that physics teachers have to battle against in the classroom is the in-built idea that students have about how objects move along circular paths. Almost everyone that comes in through the laboratory door has a committed belief in the existence of *centrifugal forces* (outward pointing).

This is perfectly understandable. Much of our common experience of circular motion seems to indicate that a force pushes you outward when turning on a curved path. For example, if you are sitting in a car that takes a corner sharply your body tends to sway outwards away from the centre of the circle.

Despite the powerful persuasion of such experiences, there is no centrifugal force pushing you outward as the car turns. In fact, the reason you move outwards is that there is insufficient force pulling you inwards.

When a car turns a corner a force must have been applied to change the direction of motion (Newton's first and second laws). A moment's thought about the way in which the direction of travel is changing ought to convince you that the force must be pulling *inwards* towards the centre of the circle (as that is the way in which the path is changing). This force is provided by the friction between the tyres and the ground. The person sitting inside the car must have a similar force acting on them if they are to follow the same path. In their case it must be provided by the friction between the seat of their pants and the car seat in which they are sitting. If this force is insufficient, then the passenger will follow a straighter path than the car. This straighter path will move them outwards relative to the car as the car turns underneath them. Eventually the side of the car or the side of the seat will come to their aid and provide an additional force to push them round with the car.

In orbit the only force acting on all the objects is gravity, and as a result they all naturally follow identical paths (unless they happen to have been knocked about inside the spacecraft). Gravity is providing the *centripetal* (inward pointing) force required to maintain the circular orbit.

The size of the force required to maintain a circular path depends on the mass of the object, the speed at which it is travelling and the radius of the circle.

- The more massive the object the greater the force required. This is a direct consequence of Newton's second law of motion.
- The faster the object is moving the greater the force required. This is also a consequence of the second law. Remember that Newton's law is best expressed by saying that the force required is equal to the rate at which the momentum is changing. Momentum is mass times velocity, so both factors influence the size of centripetal force.

When considering circular motion it is often easier to work in terms of the *angular velocity* rather than ordinary linear velocity. Angular velocity is the angle through which an object has turned divided by the time it has taken to perform this manoeuvre. In everyday life angular speeds (rates of rotation) are most often quoted in terms of the number of revolutions per minute. For example, we quote the revs at which an engine is running in rpm (revolutions per minute). Vinyl records spin on their turntables at either 45 rpm (for singles) or 33 rpm (for LPs). Compact disks can spin at rates up to 500 rpm. Physicists are not very comfortable when using terms such as revolutions (which do not make their equations look very pretty) or minutes (not a standard unit). They prefer to use *radians per second*. Those unfamiliar with the radian unit of angle can consult the box on the subject.

In physics angular speeds are quoted in radians per second (rps, or rad s^{-1}—not to be confused with rpm) and given the symbol ω. The advantage of using angular speeds is that they are independent of the distance from the centre of the circle, whereas linear speeds are not.

The Earth rotates on its axis once every 24 hours. This is an angular velocity of 2π radians in 24 hours ($24 \times 60 \times 60$ seconds $=$ 86 400 seconds) or 72 microrad s^{-1} (put that way, it doesn't sound very fast, does it?). In linear terms, however, as the Earth has a radius of 6400 km a point on the equator has to cover the circumference of a circle (40 212 km) in 24 hours, which is a speed of 465 m s^{-1}. A spacecraft orbiting the Earth once every 24 hours at a height of 35 850 km would have to move at 3.07 km s^{-1}. The angular velocity of the point on the equator and the spacecraft would be the same—but the linear speeds are rather different.

The centripetal force required to keep an object swinging in a circular path can be calculated from the simple equation:

$$F = \text{mass} \times (\text{angular speed})^2 \times \text{radius} = m\omega^2 r.$$

Radians

Most of us are familiar with the degree as a measure of angle. After all, this is the unit used on the protractor that we first came across early in our school careers. Protractors work by taking a full circle and dividing it in half (a cut across the diameter). The remaining half circle is then divided up by marking 180 equally spaced dashes along the circumference of the semicircle. When these dashes are connected up to the centre by drawing lines along the radius, the angle between two successive lines is defined to be one degree.

Have you ever stopped to think what an arbitrary choice 180 (or 360 for a full circle) is? It seems entirely natural simply because most of us have been using it since we were teenagers. Yet it is an arbitrary choice, and not an especially sensible one at that. It is not sensible as it has nothing to do with the properties of circles. There is nothing in the mathematical features of a circle to suggest that dividing it into 360 equal pieces would be a good idea. This would not matter so much if there was no sensible alternative (if you are going to make a choice between equally arbitrary alternatives then 360 is as bad as any other), but there is. Every circle has a circumference equal to 2π times its radius. This suggests to mathematicians that it would be far more natural to divide the circumference up into 2π equal pieces. The snag is that π is not a whole number (3.141 592 653 589 793 238 4...) so in practical terms a protractor could not be made to show this radian measure of angle. However, practical matters tend not to stop mathematicians. It is a simple matter to convert from degrees to radians and back again. One complete circle (360°) is 2π radians. From this it follows that 180° is π radians, 90° is $\pi/2$ radians and, by simple proportion:

$$\theta \text{ degrees} = \frac{\theta}{360} \times 2\pi \text{ radians.}$$

So, if we keep the angular rate and the radius constant, then try to put different mass objects into this orbit, they will all require different forces to act on them. In fact the size of the force will be proportional to the mass. On the other hand, if we try putting objects of identical mass into orbits with the same angular rate but at different radii, then the force required is proportional to the radius. However, there is one important point to keep in mind. Unless the force keeping the object in orbit is supplemented by engine thrust (or some other mechanism—see later), then gravity will have to do the job on its own. We cannot change the force of gravity, it is the size that it is at different distances from the Earth (or any other planet for that matter). In which case, there is a single speed which can be used by an object at a given distance from the Earth if it is to remain in orbit.

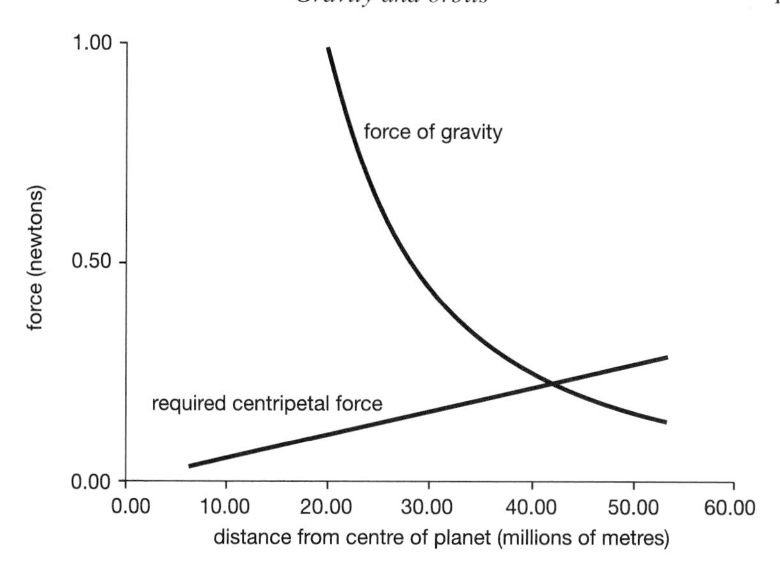

Figure 4.4 Comparing gravity with the centripetal force required to keep an object in orbit.

4.5 Gravity and orbits

Given that the force required to keep an object moving in a circle at a constant angular rate *increases* the bigger the radius of the circle, and that the force of gravity *decreases* the further you get from the planet, then it makes sense that gravity can provide the force for circular motion at a given rate at one single radius. The graph in figure 4.4 shows how the centripetal force required to keep an object circling the Earth once every 24 hours increases with distance from the centre of the planet. Also shown is the manner in which the pull of gravity decreases with distance. There is one specific place at which the two lines cross. At this distance the force of gravity provided by the planet is equal to the centripetal force required.

If you try to place an object in orbit moving at this specific angular rate but nearer than the crossover distance on the graph, then gravity will be greater than the centripetal force required. Consequently the object will fall in towards the Earth. On the other hand, if you try to place it too far out then gravity will be insufficient to hold it and it will tend to fly away. Placed at that single distance, moving at that speed, the object will continue to circle the Earth indefinitely.

The particular orbit being discussed here is very special. In terms of physics it is quite ordinary, but in terms of its usefulness to people on Earth it is unique. As far as anyone standing on the surface is concerned, any satellite placed in

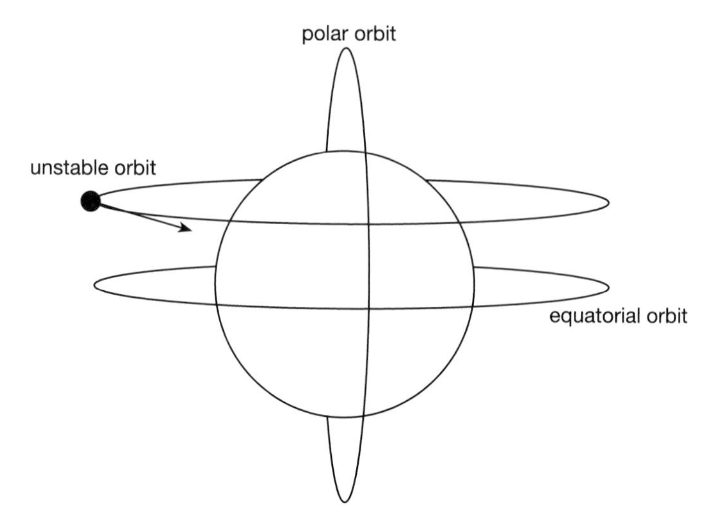

Figure 4.5 Three different orbits.

this orbit will appear to hover stationary above the ground. In fact the satellite is racing round the Earth once every 24 hours, but as this is the same time as it takes the Earth to rotate on its axis, it is simply keeping up with the ground. This is known as a *geostationary* orbit and it is the orbit of choice for communications satellites. Its big advantage is that we do not have to keep altering the aim of the communications dishes on the ground—the satellite remains 'hovering' in the sky at the same place relative to the ground. This is why the satellite TV dishes that are sprouting all over the place can be pointed in one direction and left—they are pointing to a satellite in a geostationary orbit over the equator.

Figure 4.5 shows three different possible paths round a planet. An equatorial orbit places the satellite spinning round the planet's equator and so it will be seen directly overhead by people living on this latitude. A geostationary orbit is an equatorial orbit, but it is the only one in which a satellite will turn about the Earth in the same time period that the Earth itself takes to turn about its axis.

A polar orbit has the satellite circling from one pole to another. Viewed from Earth a satellite placed in a polar orbit does not appear stationary, so that the radio dishes communicating with it have to move to track its progress across the sky. Not only is the satellite moving from north to south (or the other way), but also the world is turning from west to east beneath it. If we wait for long enough, then the satellite will fly over every part of the Earth. This makes polar orbits very handy for spy satellites as well as the more conventional research satellites looking at weather patterns, plant growth, land usage etc.

The unstable orbit shown in figure 4.5 cannot last as the force of gravity acting on the satellite is pulling it towards the centre of the planet, which is not the same as the centre of the circle in which it is trying to move. Such an orbit cannot be sustained without some other force acting (such as thrust from engines).

As I mentioned earlier, a geostationary orbit is very handy for communications satellites. So much so that it is starting to get rather crowded. Although there is physically a great deal of space between the satellites, there is a minimum distance that they must be separated by in order to prevent the signals from one overlapping with those from another by the time they reach the surface. Of course an infinite number of equatorial orbits exist, but only one geostationary orbit. At least, there is only one *unassisted* geostationary orbit.

In principle we can get a satellite turning about the Earth once every 24 hours at any distance from the surface, provided we accept that gravity will be either too great for the required centripetal force at that distance, or too small. In either case gravity must be supplemented by other means. Using thrust from engines requires a great deal of propellant and would mean that the satellite would have a finite useful time in orbit. It is far better to use a naturally occurring resource—sunlight.

A satellite equipped with large mirrors can use the pressure of sunlight reflecting off the mirrors' surface to provide an upward thrust which helps to balance the pull of gravity. In this manner a satellite could be placed in a powered geostationary orbit that was nearer to the Earth than the gravity-based one. This would require the use of enormous mirrors, but in space thin aluminium foil stretched over a light framework forms a perfectly good mirror as it does not have to survive wind, rain or other inconveniences that we experience in the atmosphere. The Russians have already launched smaller orbiting mirrors. In their case they are experimenting with the idea of reflecting sunlight to Earth to illuminate the long twilight of areas such as the Steppes. The technology would be the same.

4.6 Other orbits

In figure 4.4 I showed how both the force of gravity and the centripetal force for a geostationary orbit changed with distance from the Earth. This was just one example of an orbit. It is possible to put a mass into orbit at any distance from the centre of the Earth, provided the correct speed is chosen. Given that the orbit is not power assisted, and that the force of gravity is a fixed quantity at any radius the speed has to be chosen so that the needed centripetal force is provided by gravity.

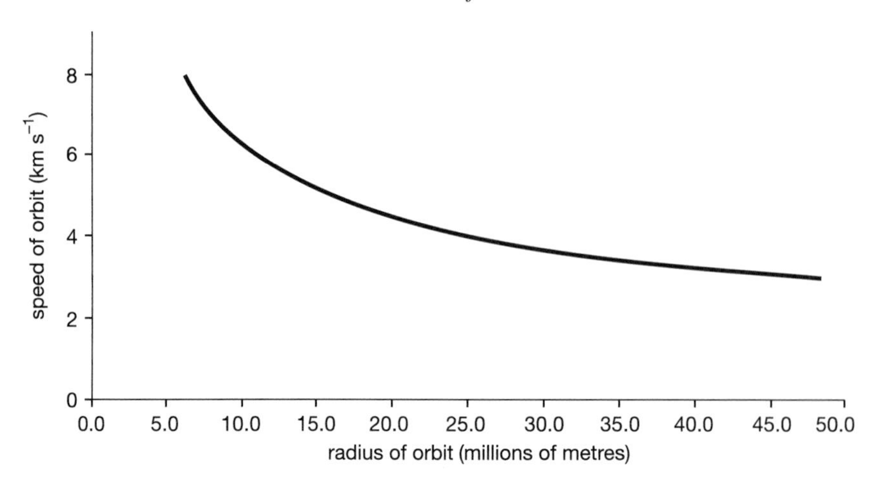

Figure 4.6 How the linear speed needed for a stable orbit decreases with distance from the centre of the Earth.

As the graph in figure 4.6 shows, the farther we are from the centre of the planet (in this case Earth) the slower the orbit has to be. This is because the force of gravity decreases with distance from the centre of the Earth[6]. This will be an important relationship to remember later in this chapter when we come to discuss how to move from one orbit to another.

4.7 Simulating gravity

With a large enough space station, or spacecraft, it is possible to use rotation to simulate the effects of gravity. As an example of how this might be done, consider an astronaut with magnetic boots standing on the inside of a rotating drum which is in space far away from any star or planet so that there is no gravitational force acting on it or its occupants (see figure 4.7).

The astronaut has a ball in her hands. While she is holding the ball, the force between her hands and the ball is providing the necessary centripetal force for the ball to match the rotation of the astronaut (her centripetal force is due to the magnetic boots[7]). At the moment she releases the ball, it no longer has any centripetal acceleration. Consequently it carries on in a straight line with the velocity it had at the moment it was released (assuming that air friction is too small to be of any concern).

Figure 4.8 illustrates what would happen to the ball. Each part of the diagram is a small period of time after the previous part. They have been moved along

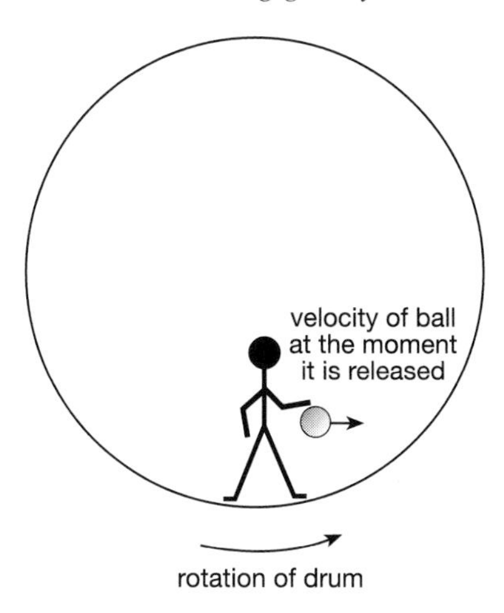

Figure 4.7 An experiment in simulating gravity.

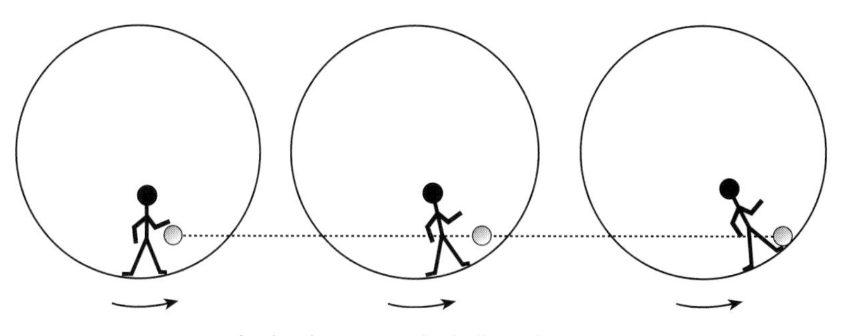

Figure 4.8 Further experiments with simulated gravity.

in a horizontal direction to make this clearer. The ball continues in a straight line as marked by the dotted line. Eventually it will reach the side of the drum (the natural tendency is to think that it will now roll back, but remember this is not a drum on Earth, there is no gravity acting downwards in this diagram!). From the astronaut's point of view it has fallen as if being pulled towards the side of the drum due to a gravitational force. When the ball hits the drum, the force of the drum on the ball will provide a centripetal acceleration so it will keep pace with the rotation from then on.

This is a simulation of gravity, it is not real gravity. A stationary ball placed precisely at the centre of the drum would stay there—there is no force pulling it to the side.

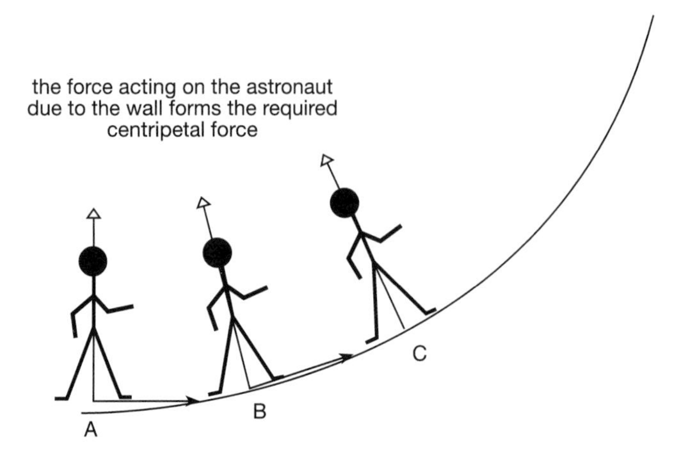

Figure 4.9 An astronaut without magnetic boots would also feel a gravity-like effect.

An interesting example of this was shown in an episode of the science fiction series *Babylon 5*. The series is set on a giant space station shaped like a cylinder rotating about an axis along its length. One of the characters was forced to jump from a train car that was running along a track mounted along the central axis (a bomb was about to go off). Although the station was spinning fast enough to simulate full Earth gravity for those living on the inside surface of the cylinder, the character did not fall towards the surface. Instead the force that was exerted on him by the train when he jumped clear had given him a forward velocity and he was slowly drifting towards the 'ground'. However this was not a safe situation as the 'ground' was rotating at a great rate beneath him. It would be like parachuting onto a moving train. The vertical speed would be nice and gentle, but the horizontal speed...

Now consider an astronaut without magnetic boots. At the point of first contact with the walls of the drum, friction would establish a forwards motion parallel to the rotation of the drum at that moment. The astronaut would then drift along at constant speed in a straight line until striking the rotating wall of the drum (i.e. from A to B in figure 4.9). At this point the impact would provide a force. Part of this force is friction and it would establish a new velocity for him. The other part is the side of the drum pushing him back (which is why he does not break through the surface of the metal). This force would give him a centripetal acceleration.

Now, in practice the astronaut would not drift across like this as the curvature of the walls would be far less. However, it does illustrate that the frictional force couples the astronaut to the drum's rotation, producing an inward-facing force that would compress his joints as he pushes against the rotating drum. Consequently he feels as if he has gravity acting on him.

The astronaut would feel as if he was on Earth when the rotation rate was sufficiently quick to produce a $1g$ centripetal acceleration at the inside edge of the drum. In practice this would need a large drum, otherwise the rotation rate would be ridiculously high.

4.8 Changing orbits

After some debate within NASA the method chosen to land a man on the Moon was selected from three viable options. The first involved sending a rocket directly to the surface, the second linking craft together in Earth orbit and the third (the one chosen) required a command module (mother ship) to dock with a lunar module in orbit round the Moon. In chapter 5 we will discuss the pros and cons of the three methods and why NASA came to the conclusion that it did.

Once that decision had been made it was immediately clear that NASA and its astronauts would have to become much more familiar with the mechanics of moving from one orbit to another so that spacecraft could rendezvous with each other. Also, the mission plan called for the spacecraft to leave the Earth's orbit and travel to the Moon, entering into lunar orbit once it arrived. All these manoeuvres would require ΔV changes brought about by firing main engines or thrusters. In order to understand how this could be done we need to discuss the shape of orbits in more detail.

So far we have been discussing circular orbits. However, exactly circular orbital paths are quite rare. Most objects orbit on elliptical paths.

Mathematically an ellipse is described by fixing two points in space called the *foci* (see figure 4.10). Any point on the path can be connected to these points by straight lines (radial lines). An ellipse is the only path for which the sum of the distances to each focus remains the same as we move along the path.

Johannes Kepler was the first person to realize that the planets orbit the Sun in just such paths. He discovered this by painstakingly tracking the orbit of Mars (using observations produced by Tycho Brahe) and trying to match it to a mathematical shape. Some years later Newton, using his newly discovered law of gravity, proved that all objects must orbit in elliptical paths.

In the case of a planet orbiting the Sun, or a satellite in orbit round the Earth, one of the foci of the ellipse must be the centre of gravity of the attracting

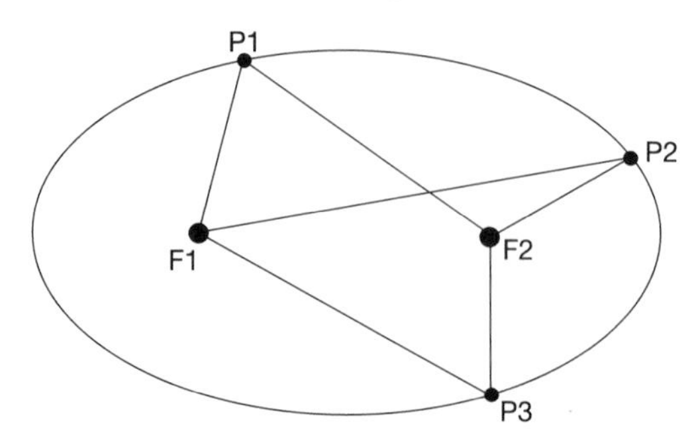

Figure 4.10 An elliptical path. The two foci are marked as F1 and F2. At each of the three points illustrated (P1–P3), adding together the lengths of the lines connecting the point to the foci always gives the same answer.

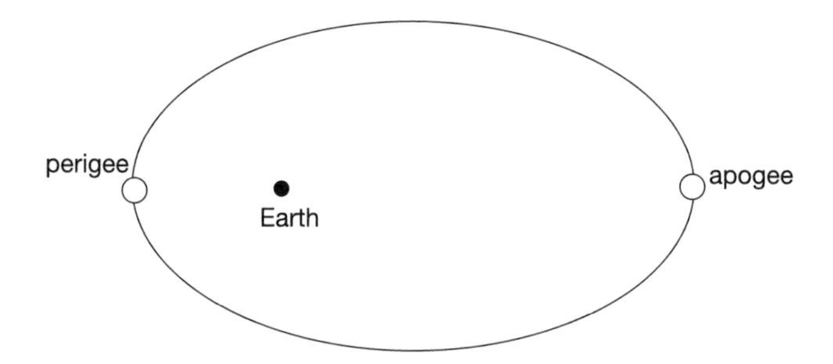

Figure 4.11 An elliptical orbit about the Earth has a point of lowest altitude (called the perigee) and one of highest altitude (called the apogee).

object[8]. The other focus is simply a spot in space—there does not have to be anything there.

As a satellite loops about the Earth on an elliptical path, it passes through a point where it is nearest to the ground (the *perigee* of the orbit) and, at the opposite end of a line through the centre of gravity of the Earth, a point where it is furthest from the ground (the *apogee*), see figure 4.11. The greater the distance between the foci the more *eccentric* the orbit and the greater the difference between apogee and perigee. A circular orbit has both foci in the same place (the centre of the circle), and no distinguishable apogee and perigee. In practice all the planetary orbits (with the exception of Pluto's) are so nearly circular that a casual glance at their paths drawn on a scale diagram would be unable to distinguish them from exact circles.

Another factor that has to be taken into consideration when plotting the path of a satellite or spacecraft is the speed at which it is moving in the orbit. On a circular orbit the speed is constant. However, with an elliptical path the speed

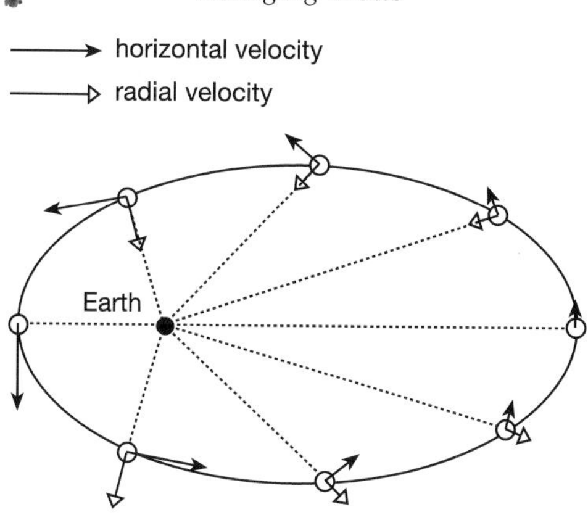

Figure 4.12 The horizontal and radial components of a spacecraft's velocity in an elliptical orbit.

varies from a maximum at perigee to a minimum at apogee. In the case of a circular orbit the velocity is always directed along a tangent[9] to the path; for an ellipse this is not so. The easiest way of tracking the change in velocity is to consider the motion as being split into two parts—along the line connecting the spacecraft to the centre of gravity of the planet (radial motion), and along a line at 90° to this one (horizontal motion[10]).

As can be seen in figure 4.12, the only two points at which the velocity is purely horizontal are at apogee and perigee. A spacecraft passing through its perigee point is moving faster than an equivalent craft on a circular orbit of the same radius. Consequently the force of gravity at perigee is not enough to hold it in a circular orbit. As it moves away from perigee it also tends to move away from the Earth—it gains a radial component of velocity pointing away from the centre of the Earth. However, it is not moving fast enough to totally escape the pull of the Earth's gravitational field. It loops away slowing down as it goes (just as a rock thrown into the air will slow down to rest before falling back). At apogee its radial velocity reaches zero, so it once again has a purely horizontal velocity. However, now it is moving too slowly for a circular orbit at this radius, so the gravitational force pulls it in. Of course it does not fall directly towards the Earth as it has not lost its horizontal velocity. It loops back towards perigee. The pattern repeats itself unless some other force takes a hand—such as air resistance, if perigee is low enough for the spacecraft to skim the atmosphere, or the spacecraft's engines burning.

If a spacecraft in an elliptical orbit were to light up its engines, then the change in orbital path resulting would depend on the ΔV of the burn, the direction of thrust applied and the point in the orbit at which the burn took place. All of these are complicated factors and the general effect on an orbit requires careful calculation. However, it is possible to make some simplifying assumptions that will enable us to see the effects of some types of burn. Initially we will assume that:

- the burn takes place either at perigee or apogee;
- the thrust direction is along the line of flight;
- the burn duration is very short compared to the orbital period.

If the first two assumptions are true then the effect of the burn is to raise or lower the horizontal velocity of the spacecraft at the points where it only has a horizontal velocity. The third assumption is sometimes known as the impulse approximation—that the ΔV produced by the burn arrived instantaneously and at a single well defined point on the orbital path. Of course, in practice the ΔV is spread over a burn duration and so the spacecraft moves along the orbit while this is happening—which has an effect on the final orbital change achieved.

For example, consider a spacecraft flying in an elliptical orbit and burning its engines at the moment it reaches perigee. The subsequent path it takes will depend on whether the ΔV increased or decreased the velocity.

Decreasing the velocity will mean that the spacecraft moves away from perigee less quickly. Consequently the centripetal force that would be required to keep it in a precisely circular orbit is lower and so closer to the actual force that the Earth's gravity can provide. The path is more nearly circular. In fact, with

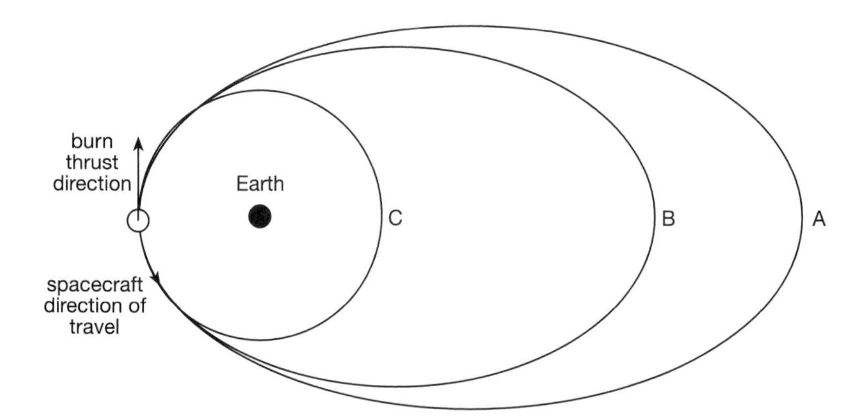

Figure 4.13 Firing the engines at perigee with a thrust direction that is against that of the horizontal velocity will slow the spacecraft down. (Remember that the direction of thrust is opposite to the direction in which the exhaust is emitted.) As a result path A will be turned into a less eccentric orbit such as B, or even into a circular orbit (C) if the ΔV is correct.

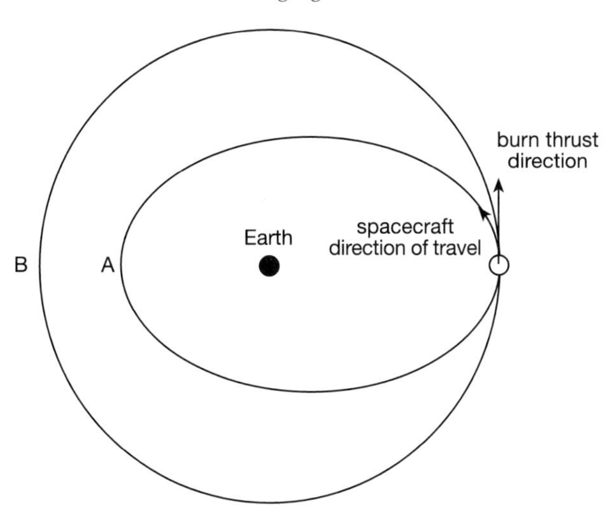

Figure 4.14 A burn performed at apogee of an elliptical path (such as A) that increases the speed of a spacecraft will tend to raise the height of perigee. With sufficient ΔV the height of perigee can be lifted to being equal to that of apogee—i.e. the orbit has become circular.

a correct burn the spacecraft can lose sufficient velocity to change from the elliptical path to a circular one with a radius equal to the radius at perigee on the old ellipse. This is the process of *circularizing* an orbit.

Increasing the velocity will have the opposite effect. The spacecraft enters into a more eccentric orbit, however the height and position of perigee remains the same. Apogee moves further away.

The same two types of horizontal burn can be applied at the point of apogee. If the resulting ΔV increases the velocity then this will tend to circularize the orbit to a radius equal to that of the apogee. On the other hand, slowing the spacecraft down by a horizontal burn at apogee will tend to make the orbit more elliptical.

The other simple type of burn is a radial one. In this case the spacecraft is oriented so that the engines thrust along the line connecting the craft with the centre of the planet. Radial burns can also be used to circularize an elliptical orbit. At any point on the ellipse between apogee and perigee a spacecraft will have both a horizontal and a radial velocity. If a burn is timed to have a ΔV equal to the radial velocity at the moment of burn, and directed in the opposite direction, then the radial velocity can be cancelled out. This will convert the orbit into a circular one—an orbit with no radial velocity. Similarly, radial burns can be used to convert circular orbits into elliptical ones. Unlike the case of horizontal burns which convert the position of the spacecraft into the apogee or perigee of the new orbit, a radial burn will move the craft onto an orbit somewhere between apogee and perigee, depending on the ΔV achieved.

Hohmann transfers

In 1925 a German engineer, Dr Walter Hohmann, showed mathematically that the way to transfer between two circular orbits of different radius using the least propellant was by an intermediate elliptical orbit. This technique is now known as a *Hohmann transfer*.

Let us say that a spacecraft starts in a low circular orbit and the occupants wish to move it into a higher circular orbit (say to rendezvous with another spacecraft or satellite). They start by performing a burn that accelerates their horizontal velocity. This changes them from a circular orbit to an elliptical one. The perigee of their new orbit is the point at which the burn took place (remember we are still using the impulse approximation). The ΔV achieved by the burn will decide the height of the apogee which they will reach on the opposite side of the orbit to the burn point.

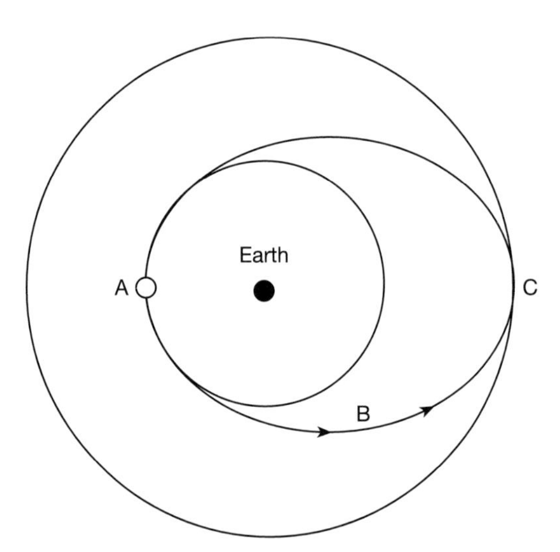

Figure 4.15 A Hohmann transfer between a low orbit and a higher orbit. A spacecraft at point A carries out a burn that increases its horizontal velocity. The burn point then becomes the perigee of a Hohmann transfer orbit. The spacecraft coasts along the orbit (B) until reaching the apogee at C which intersects with the required higher orbit. Another burn at this point can accelerate the craft, placing it on the higher circular orbit.

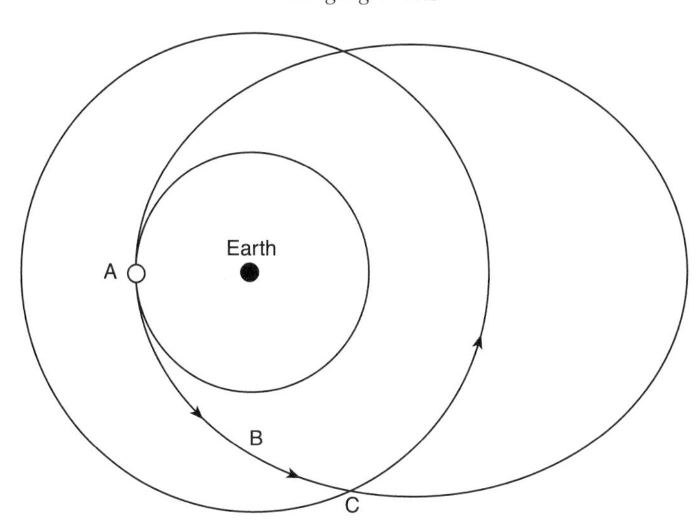

Figure 4.16 This transfer path is faster than the Hohmann ellipse. However, the manoeuvre requires more propellant—at A to put it on an ellipse with a much higher apogee, and at C a combination of a horizontal and radial burn to deflect the spacecraft onto the circular path.

The astronauts select the ΔV that places the apogee of the new orbit at the same height as the circular orbit they are aiming for.

When they arrive at apogee, they must burn again to circularize their path. Slightly strangely, they again have to burn so as to accelerate their spacecraft. With the correct ΔV from this second burn they will settle into their new circular orbit.

At first sight it seems wrong that the way to move from a low orbit to a high one is to make two *accelerating* burns. After all, the high orbit will require a lower speed for stability than the low orbit! However, as the spacecraft coasts up to the higher orbit it will gradually lose speed (as a ball thrown in the air does) until the point when it reaches the apogee of the transfer orbit. There it is moving too slowly to be on a circular path (after all, its motion is not circular—it is on the elliptical transfer orbit!). At this point another burn is needed to bring it up to the required speed.

It is also possible to transfer from a high orbit to a lower one using a Hohmann transfer with two burns that slow the craft down.

Of course, the Hohmann transfer is not the only orbit that can connect two others. It just happens to be the one that uses the least propellant. It would be possible to burn the engines for a greater ΔV to place the spacecraft on an elliptical path that had an apogee much higher than the required final orbit. This has the advantage of cutting down the transfer time, but at the expense of a greater use of propellant.

4.9 Flying to the Moon

There is a film, made many years ago, composed entirely of clips from the Apollo missions. Called *For all Mankind*, it features commentary from the astronauts and so gives some idea of what they were thinking and feeling during the missions. The title of the film echoes the inscription on a plaque attached to one of the legs of the LM used in the Apollo 11 landing:

'Here men from the planet Earth first set foot on the Moon, July 1969 AD. We came in peace for all mankind'

One of the most interesting parts of the film deals with the long wait on the launch pad. The three astronauts were strapped into the command module about 90 minutes before the launch while the mission controllers went through the pre-flight checks, brought the propellant tanks up to pressure and other tasks. While there were procedures for the three of them to carry out, some time (especially during a 'hold' in the countdown due to a fault) was spent simply waiting. Lying on one's back strapped to a 2.7 million kg bomb and twiddling your thumbs while waiting for it to go off is not the most relaxing situation that can be imagined.

In the film there is a nice sequence shot out of one of the command module windows. The Moon in a cloudless blue sky is visible directly above the spacecraft. The commentary runs:

> *'I know they are doing their job right because the Moon is right straight ahead and that's where we are pointed, an' they're just going to launch us right straight to this thing.'*

The business of flying to the Moon is not as simple as pointing the craft in the right direction and firing the motors. Various obvious factors have to be taken into account—such as the fact that the Moon is a moving target and so you have to aim for where it will be when you arrive, not where it is when you launch. However, there are some less than obvious factors to consider as well—such as the angle of the sunlight on the lunar surface when you arrive, so that there is good contrast and visibility for the landing.

Once the systems had been thoroughly checked out in Earth orbit, the Saturn V third stage fired again, placing the craft on a large elliptical orbit out towards the

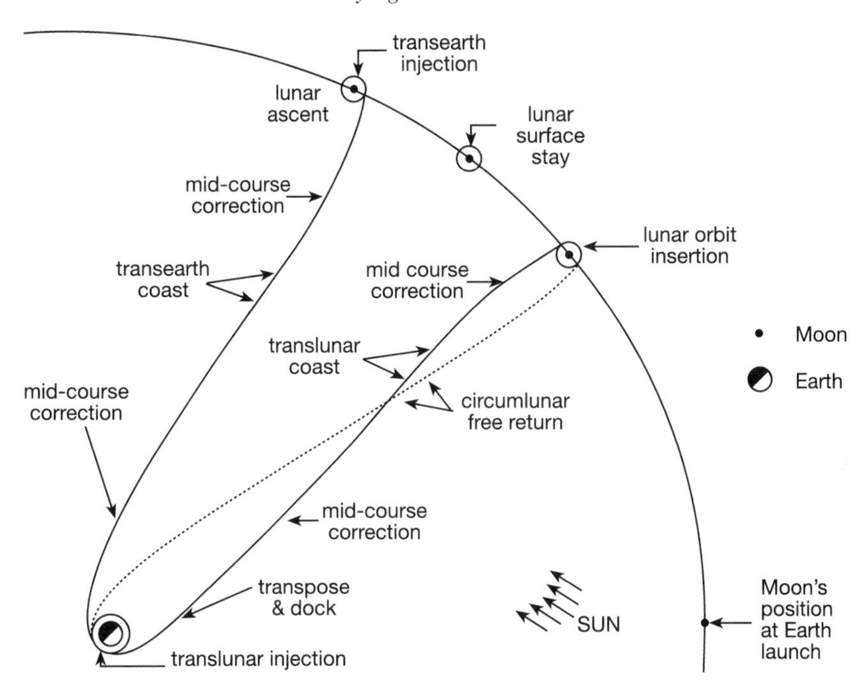

Figure 4.17 The Apollo 11 lunar trajectory (adapted from a NASA diagram).

Moon. Two possible mid-course corrections (to fine-tune the trajectory) using the service module's engine were budgeted for in that engine's ΔV allowance.

As Apollo got closer to its destination the Moon's gravity started to swing the path round towards it. If the service module engine failed to fire then the Moon would loop the craft round onto a free return to Earth (some missions departed from the free return trajectory on the way out so that they could explore other parts of the Moon's surface).

As Apollo was pulled round behind the Moon the service module engines fired to slow the craft into an elliptical orbit about the Moon. On Apollo 11 this first *lunar orbit insertion* burn (LOI in NASA speak) had a ΔV of -892 m s^{-1} and placed the craft into an elliptical orbit with an apocynthion[11] of 315 km and a pericynthion[12] of 111 km (both measurements taken to the surface of the Moon, not the centre). About four hours and two orbits later the service module engine was fired again to circularize the orbit (ΔV -48 m s^{-1}) at 100 km by 120 km. Interestingly this was a very well designed orbit as the variations in the Moon's gravity (due to mascons etc) would make it totally circular by the time the LM was due to dock with the command module again after the landing.

The return trajectory was essentially the same as that on the outward-bound trip. With the craft now considerably lighter (no lunar module) and the Moon's gravity being less than that of the Earth, the service module engine was sufficient to accelerate the craft out of lunar orbit and onto an Earth-bound trajectory.

4.10 Trajectories to Mars

At first consideration the journey to Mars is daunting in its length. Apollo travelled at about 1.5 km s^{-1} on the way to the Moon[13]. At that rate the trip to Mars would be 8.5 *years* away. There is no current technology that could build a craft to sustain life for that period of time.

What hope is there then for a Mars mission without some dramatic improvement in technology to increase the speed of our vessel?

In fact the situation is not that bad. Firstly, the average speed to the Moon is a lot less than might have been achieved with the Saturn V third stage. Properly fuelled the stage would have been capable of boosting Apollo to 4.5 km s^{-1}—at which speed the trip would have lasted just over a day. However, the service module's engine would not have been powerful enough to brake Apollo in orbit about the Moon. The Moon's gravity helps the manoeuvre, but even so the 1.5 km s^{-1} average speed used was about the highest practical. A Mars mission does not have the same limitations. For one thing, Mars' gravity is considerably greater than that of the Moon, so it is capable of pulling in a faster-moving spacecraft. For another, Mars has a thin atmosphere—the Moon has none.

The advantage of an atmosphere is that it can be used to help slow the spacecraft down. The technique is called *aerobraking* and it has already been successfully used on both the Mars Global Surveyor and Magellan missions (the latter used radar to map the surface of Venus). An aerobraking manoeuvre dips the spacecraft into the upper layers of the atmosphere and uses the friction produced to help slow it down. Using aerobraking means that less propellant has to be carried to be used to slow the spacecraft down, and so it can travel with a much greater velocity.

There is another factor that helps get a manned mission to Mars sooner than you might have thought. When a spacecraft leaves Earth's orbit heading for Mars it benefits from the additional speed with which the Earth is moving. Remember that the Earth is itself moving round the Sun at something like 30 km s^{-1}. This is rather like throwing a ball from the front of a moving train. A craft launched at 3 km s^{-1} would move away from Earth's vicinity at a speed of

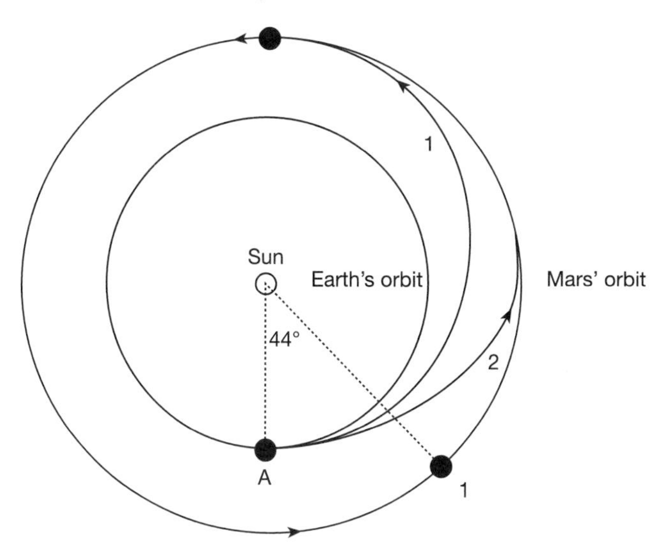

Figure 4.18 Two possible trajectories for a Mars mission. Both Earth and Mars orbit in an anticlockwise direction if the orbits are viewed from above (as in this diagram). Mars takes roughly two years to orbit the Sun, so in the time that Earth moves half way round the Sun, Mars advances a quarter of its orbit.

33 km s^{-1} relative to the Sun[14]. Once various factors are accounted for, such as the slowing down due to the pull of the Earth's and the Sun's gravitational force, we end up with a Hohmann transfer time of about 250 days—which is far more manageable.

This extra boost does not even significantly affect the braking at the other end. After all, Mars is also moving round the Sun at 24 km s^{-1} and when the spacecraft arrived it would be travelling nearly parallel to Mars. That gives a relative velocity of 3 km s^{-1} as by then the craft would have slowed to more like 21 km s^{-1}.

Taking all these factors together implies that, surprisingly, a mission to Mars can be flown from low Earth orbit with a ΔV that is actually *less* than that required to get to the surface of the Moon (see figure 3.10). Two possible routes to Mars are illustrated in figure 4.18.

The Hohmann transfer orbit (1) to Mars uses the least propellant and takes about 258 days (see the next page). As Mars takes 687 days to complete its orbit about the Sun, the transfer time on this trajectory represents 258/687ths of an orbit. In this time Mars will move 135° out of the full 360° around the Sun. As we discussed earlier, a Hohmann transfer will arrive at the higher orbit a full

180° after the starting point. To ensure that Mars is there when the craft arrives (clearly desirable), the launch must take place when Mars is $(180° - 135°) = 44°$ ahead of the starting point (e.g. position 1 in figure 4.18).

It is possible to reduce the transfer time by using more propellant and moving onto an elliptical orbit with a higher aphelion[15]—such as trajectory 2 in figure 4.18.

Unfortunately, due to the fact that Earth's and Mars' orbits do not lie in exactly the same plane about the Sun (there is a 2° difference between them), exact Hohmann transfer orbits do not exist. However, there are two classes of orbit that are very close to the ideal Hohmann transfer. A type 1 orbit takes more like 200 days to reach Mars, but due to the faster speed would require more propellant to slow it into a Martian orbit. A type 2 orbit takes 300-odd days to reach Mars, but requires less propellant when it gets there. The choice of type 1 or type 2 largely depends on the mass of the craft being sent. Low-mass craft, and those sent to fly past Mars rather than to orbit, travel the type 1 route, which is faster, and their lower mass helps compensate for the greater mass of propellant needed. Heavier craft take the type 2 route. Of the recent missions to Mars, *Pathfinder* took a type 2 flight and the *Mars Global Surveyor* took type 1.

One perceived problem with a Hohmann-type transfer to Mars for a manned flight is connected with the return phase of the mission. As the relative position of the Earth and Mars are crucial, it is not possible to set off back home at any moment. If the crew want to return on a minimum-propellant mission, then they must wait for a 'launch window' to open. This will involve a stay on Mars of between 300 and 500 days. Some analysts regard this as being far too long a time to survive in a hostile environment (however there are some more radical views, as discussed in chapter 8). It is possible to use a much faster transfer orbit that reaches Mars in about 60 days and then a low-propellant return window opens almost immediately on arrival. The problem with this is that such an outward-bound mission would require a very large ΔV and so the payload would be much reduced.

A more imaginative solution is shown in figure 4.19. This trajectory initially starts in the wrong direction. The craft falls in towards the Sun, gaining speed as it does so, due to the pull of the Sun's gravity. As it loops behind the Sun it encounters Venus in its orbit. Venus' gravitational pull swings it round into a new orbit out towards Mars. In this way the craft has gained a considerable ΔV without the use of propellant. In figure 4.19 V1 marks the position of Venus at launch. V2 is the position of Venus when the craft encounters it and M1 the position of Mars on the launch date. Such trajectories have been used already with unmanned interplanetary probes.

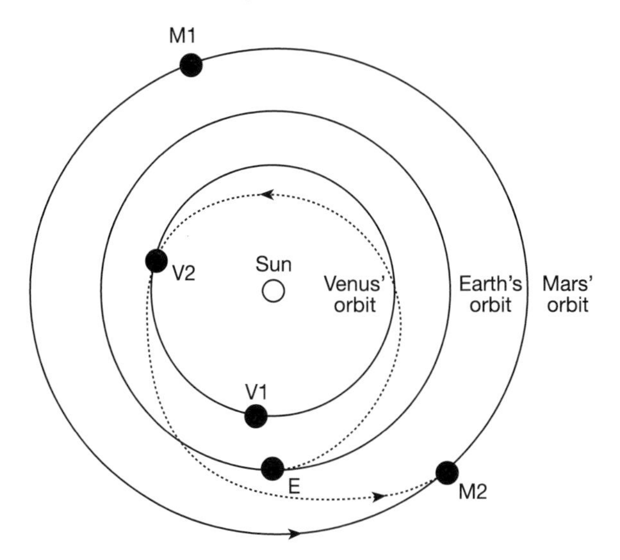

Figure 4.19 One solution to the problem of return windows is to use a Venus-gravity-assisted trajectory.

However, this is a far from ideal solution. The orbit has been termed a 'Venus fryby' by those that believe that the proximity to the Sun on this path has severe implications both for the thermal shielding of the craft and also the protection of the crew from the greater intensity of ionizing radiation[16].

We might also like to consider sending the crew on a free return trajectory, along which the gravitational pull of Mars would loop their spacecraft round onto a path that automatically returns to Earth. This would be a considerable safety advantage should the engines fail. It might then be possible to use another craft to rescue the crew once they were nearer to Earth, should their vehicle also be incapable of re-entry into Earth's atmosphere.

We will take up more of the various issues involved in planning a mission to Mars in chapter 8.

4.11 Space stations

The idea of an orbiting construction in which human beings could live and work for extended periods has captured the imagination since the earliest thoughts about space travel. Now, as the century draws to a close, the first permanently manned station is being constructed. However, this is by no means the first-ever space station. There is a long history of stations in Earth orbit that have met with varying degrees of success.

Salyut

Having been beaten to the Moon, the Soviets set their minds further afield. They began a series of space station missions with research and military objectives. Undoubtedly part of the thinking behind this was to prepare for long-duration space flights (e.g. to Mars) by looking into the effects of zero g on human beings. There was also research being carried out on the ground into how to sustain a small biosystem for long periods of time.

Although all the early Russian space stations went under the name of *Salyut*, it is clear now that there were two distinct sets of missions flown, with the *Almaz* project having military objectives.

- Salyut 1 was launched on 19 April 1971 (the first space station ever to be put into orbit) and was manned from 6 June to 29 June of that year. The first crew members were launched only four days after the station had entered orbit, but they were unable to enter the station even though their Soyuz spacecraft had docked successfully. The second flight was more successful—the crew managed to enter the station and carry out a 24 day mission including astronomical, biological and Earth observation research. However, the mission was far from successful overall in that the three crew members died during the return to Earth. A valve had opened, letting the atmosphere out of the capsule during re-entry. Although the craft landed successfully on the ground (as was the Russian habit, rather than a splashdown) the crew were found to be dead when the recovery team arrived.
- Salyut 2 was the first of three Almaz-type stations. It was launched on 3 April 1973, but a control malfunction caused the station to start tumbling one week into the mission. It broke apart and burned up in the atmosphere on 28 May.
- Salyut 3—the second Almaz station—flew on 25 June 1974 and was manned between 3 July and 19 July 1974. The crew of two Soviet Air Force cosmonauts used cameras and TV systems to carry out photographic reconnaissance. A second flight to the station was aborted when the crew was unable to dock with the station.
- Salyut 4—this was a civilian station launched on 26 December 1974. Two successful missions were flown (29 and 63 days respectively), although the middle one of three intended crew launches had to be aborted shortly after launch.

- Salyut 5—the primary purpose of this Almaz-type station was military, although quite a bit of scientific research was also carried out on board. The first mission (49 days) docked on 6 July 1976 after the 22 June launch of the station. Once again the second planned mission was aborted, this time due to another docking failure. The third mission ran from 7 to 25 February 1977.
- Salyut 6—the first of the second generation of Soviet space stations. Up to this point, the Salyut had only been capable of docking with one Soyuz capsule, restricting the number of crew members and the length of stay on the station. Salyut 6 had two docking ports so that the crew could be rotated and re-supplied by automatic vehicles. The station was in orbit for five years (29 September 1977 to 29 July 1982), during which it was manned for 676 days. The record mission length was set at 185 days. There were also many smaller-duration flights, including some that carried international crew members.
- Salyut 7—launched in April 1982. This was a very similar station to Salyut 6. It was manned for 800 days during its nine-year orbital lifetime. It overlapped with the construction of Mir and during the first flight to that later station some equipment was transferred over from Salyut 7.

Skylab

The first American space station arose out of the premature end of the Apollo programme. Having beaten the Soviet Union to the Moon, America was not about to surrender Earth orbit to them and their Salyut space stations. By using the large fuel tank of a Saturn V third stage as a basis, the Skylab space station could be built comparatively cheaply. And it would be bigger than any Soviet station launched to that date. Skylab consisted of four distinct sections.

1. The orbital workshop—this was the section converted from a Saturn V third stage (see figure 4.20). It contained work areas, sleeping quarters, kitchen, bathroom and even a zero *g* shower unit. It was divided into two decks with a total length of 14.6 m and a diameter of 6.6 m.
2. The airlock module—this was a 5.3 m long tube 3 m in diameter which contained not just the airlock through which crews would enter but also electrical power and communications systems.
3. The Apollo telescope mount—this was one of the most important instruments on the station. One of the prime research goals of Skylab was to observe the Sun. This instrument contained eight solar telescopes for that job and was powered by a set of four solar panels.
4. The multiple docking adapter—this contained two docking ports in case an emergency rescue mission had to mounted. In such circumstances (e.g. the Apollo that carried the crew into orbit malfunctioning) a special

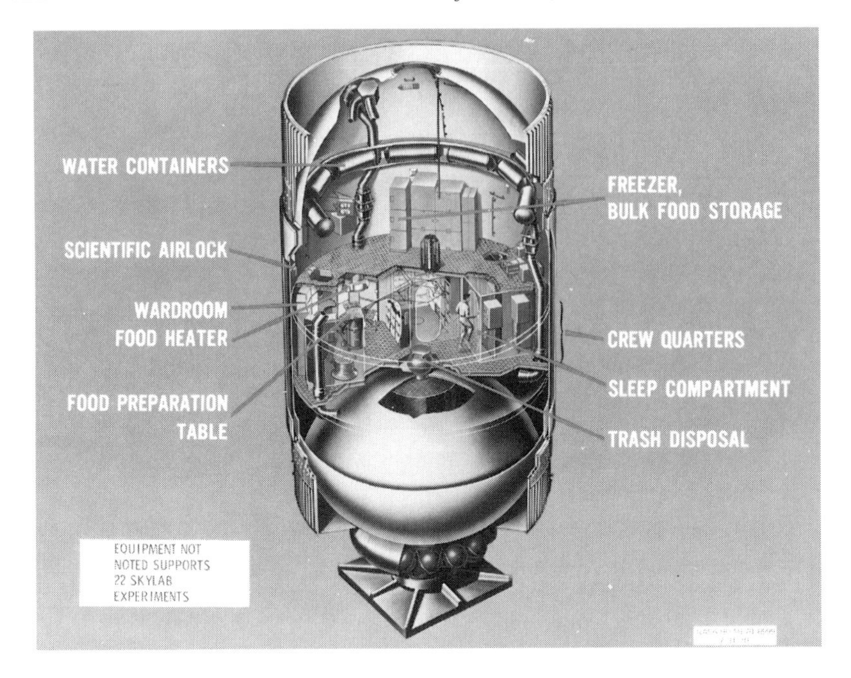

Figure 4.20 The Skylab orbital space station workshop section.

configuration could be used in the command module with two extra couches mounted on the floor underneath the standard fitting units.

The station was launched on 14 May 1973 using the first two stages of a redundant Saturn V left over from Apollo. However, the launch was not a smooth one. As the rocket accelerated and passed 7.6 km altitude, the air resistance tore at the cylindrical metal shield designed to protect the station from small meteorites and give it some insulation from the Sun's rays. The shield hung from an aluminium strap which got tangled in one of the solar panels that was folded into the structure during launch. As the rocket staged the shield was finally ripped away from the rocket, taking the solar panel with it.

The rest of the launch proceeded normally and the station made it into orbit. However, without its shield the temperatures inside rose to 52 °C—high enough to spoil the food, photographic film and experiments packaged inside. Furthermore, although the Apollo telescope mount had deployed properly and extended its solar panels, they could not provide enough power to compensate for one of the main solar panels having been torn away during launch and the other being jammed shut. The station was seriously short of power.

Figure 4.21 A view of the Skylab station from the command module used in the third and final mission. The Apollo telescope mount can be seen at the centre of a cross of four solar panels. Also the improvised heat shield covering the main compartment can be seen over the top of the station.

On 25 May the first Skylab crew took off in an Apollo CSM combination atop a Saturn 1B rocket. Their main job would be to try and repair the crippled station. The mission commander, Pete Conrad, had already walked on the Moon during Apollo 12. However, he came to think of his repair mission to Skylab as the pinnacle of his career as an astronaut.

The crew's first job was to repair the docking mechanism which was refusing to latch. Then the crew could enter the station and cover part of the main body with a Mylar heat shield. Once this had been done the temperatures inside dropped to the point at which the crew could work without space suits[17]. Next they moved on to freeing the jammed solar panel, which they did with a pair of tools that resembled a set of long-handled tree pruning shears and a crowbar.

The first Skylab mission lasted just over 28 days. The second crew lifted off on 28 July of the same year and remained on board for nearly $59\frac{1}{2}$ days. The final crew (16 November 1973 to 8 February 1974) stayed the longest—84 days. One highlight of their mission was the opportunity to photograph a comet named Kohoutek from outside Earth's atmosphere[18].

The three Skylab crews spent a combined time of over 3000 hours conducting scientific experiments in Earth orbit. They returned with over 182 000 pictures of the Sun and more than 40 000 Earth surface pictures. In addition, data from Earth observations were stored on magnetic tape and brought back to Earth—some 73 km of it!

Further missions to Skylab were planned for the space shuttle which was starting its development at the time. However, Skylab's life was cut short when it re-entered Earth's atmosphere and burned up over Australia on 11 July 1979. This had not been a miscalculation on the planners' part. During the late 1970s the Sun went through a period of intense activity on its surface. Consequently the amount of radiation striking the Earth's upper atmosphere dramatically increased. This always has the effect of expanding the Earth's atmosphere upwards. The increased drag that this produced on Skylab is what brought the station down prematurely.

Mir

Undoubtedly the most successful, and controversial, space station ever constructed is the Russian *Mir*. The name can be translated in several ways. The official version is 'peace', but the word can also mean 'commune' or 'village'. Mir is in a circular orbit at an altitude of 400 km, taking approximately 92 minutes to make one revolution around the Earth. At first the crews used to throw trash out of the airlock, but they have been dissuaded from doing that as the international space station will be in a similar orbit.

The first component of the station was launched on 20 February 1986, so it had been in orbit for twelve years as of December 1998. From the outset Mir was designed to be a modular space station. This improved on the earlier Salyut stations by moving most of the scientific experiments and equipment out of the central habitation module into specialist modules that coupled to the station.

The first module (named Mir) is the core of the whole station. It comprises the operations zone and the living zone, the latter being where the crew each have their own cabin (containing a chair, a sleeping bag and a window) and also where the galley, shower, toilet and sink are installed. The operations zone is the control area for the whole station. From here the crew can monitor most of the station's systems as well as control the manoeuvring engines. The Mir module is 13 m long and has a maximum diameter a little less than 4 m. In an attempt to make the crew feel more at home, all the habitation parts of Mir have a carpeted 'floor', coloured walls and a white fluorescent-tube-lit ceiling. Although 'up' and 'down' are simply relative terms in zero g, there are psychological advantages in this arrangement for long-term habitation.

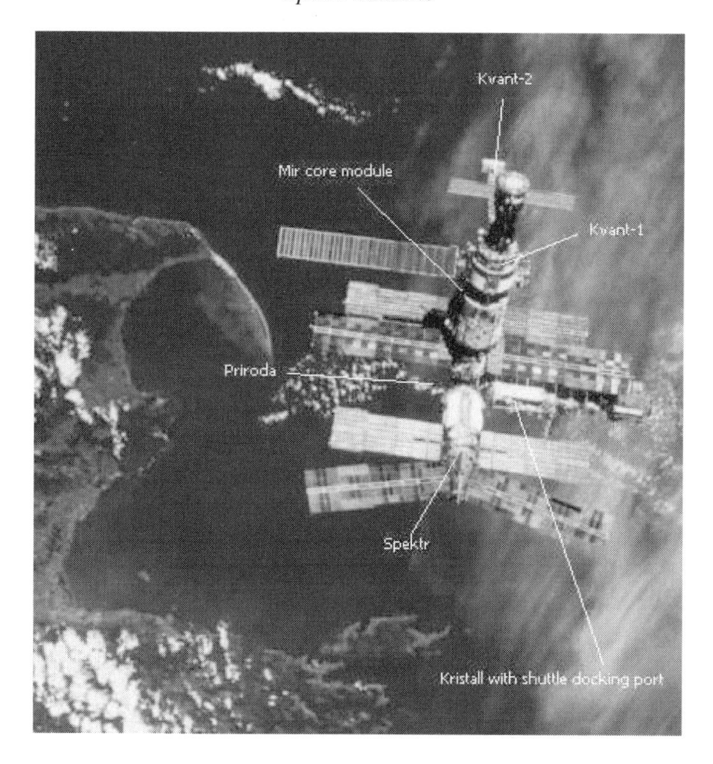

Figure 4.22 The Mir space station as photographed from an arriving space shuttle. This picture was taken before the Priroda module arrived in 1996, but its location is indicated.

In order to aid the assembly of the modular station, Mir is equipped with six docking ports. Two are at either end of the main module's long axis and four more are arranged at 90° intervals about the circumference at one end. Currently there are five modules attached to Mir, listed below in order of attachment to the main module. The best way of describing Mir is to imagine a line down the axis of the main module. If we call this the X axis, with plus X running away from the aft docking port, then there are Y and Z axes at 90° forming a right handed coordinate system (like the corner of a room).

- Kvant-1 is attached to the aft docking port of Mir. This module was designed for astrophysics research. The module is 5.8 m long with a maximum diameter of 4.3 m. It contains equipment for measuring the spectrum and x-ray emissions of such exotic objects as active galaxies, quasars and neutron stars—which is much more easily done without the absorption of the Earth's atmosphere.
- Kvant-2 (12.2 m by 4.3 m) is a scientific module that also contains the EVA airlock. Its main research areas are biotechnology and Earth observation.

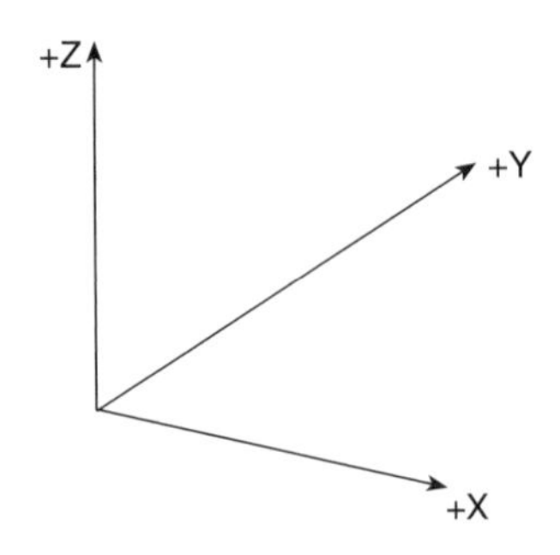

Figure 4.23 A right-handed coordinate system.

This module is linked to Mir at the +Y docking port at the opposite end to Kvant-1.

- Kristall (13.7 m by 4.3 m, connected to the −Z docking port) is a module dedicated to technological research into materials (such as semiconductor production). Some biological research is also carried out—the module contains a greenhouse used to investigate plant cultivation in zero *g* (a vital area for long-term space travel). Kristall also contains the docking port needed to mate an American space shuttle with the station. This was attached to Kristall in 1985 to add a 'spacer' to allow the docked shuttle to clear Mir's solar panels—otherwise the module would have had to be moved to the end of Mir to accommodate the shuttle every time a docking was required.
- Spektr (14.1 m by 4.3 m, connected to the −Y docking port) is another module equipped for atmospheric and Earth-surface studies.
- Priroda (13 m by 4.3 m, connected to the +Z docking port) is used for remote sensing, i.e. the study of Earth's surface and atmosphere using infra-red and various other detectors. Such research can help in the study of global warming (by the migration of crops, for example) and the ozone and aerosol constituents of the atmosphere.

Although the modules can be moved about from docking port to docking port (and have been as various new modules arrived), Mir is now in its final configuration. On 27 August 1999 what will probably turn out to be the last crew left the station. The station's owners, the Energiya space corporation, have attempted to raise $100 million to keep the station's orbit stable, but currently without success. It is likely that Mir will burn up in Earth's atmosphere sometime in 2000.

The International Space Station (ISS)

For some while NASA has been dreaming of a permanently manned space station much larger than anything attempted before. After numerous delays

Figure 4.24 An artist's impression of the completed International Space Station (as in 2004).

Figure 4.25 The first two modules of the ISS assembled in Earth orbit.

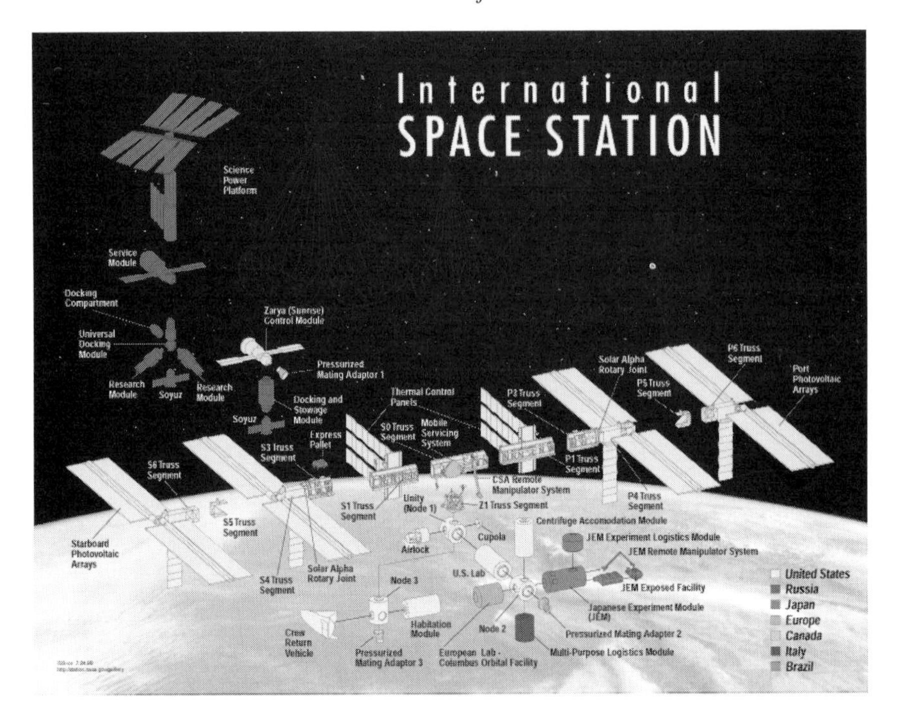

Figure 4.26 A breakdown of the various components that go to make up the ISS.

in the planning and execution of the project, the first module of the ISS was launched from Russia on 20 November 1998.

This module, named *Zarya*, is designed to provide propulsion control for the completed station and electrical power for the modules during the early stages of construction (before the main solar cells are deployed). It will also provide fuel storage.

The launch of Zarya was followed on 3 December 1998 by a space-shuttle mission (STS-88) to mate this module with the American module *Unity*. This module has six connecting ports for various future modules. It also provides a docking facility for later shuttle missions.

Over time more and more modules will be launched which have been constructed by various countries in the consortium responsible. The completed ISS will be 88.4 m long and 110 m across with modules contributed by the USA, Russia, Japan, Europe, Canada, Italy and Brazil (see figure 4.26). The construction schedule is outlined in table 4.1. The information in this table is taken from the June 1999 assembly sequence review. Further information can be found at http://spaceflight.nasa.gov/station/assembly/flights/chron.html. In addition to

Table 4.1 ISS assembly sequence: June 1999 planning reference.

Date	Flight	Launch vehicle	Element(s)
20 Nov 1998	1A/R	Russian Proton	• Zarya Control Module (Functional Cargo Block—FGB)
4 Dec 1998	2A	US Orbiter STS-88	• Unity Node (1 Stowage Rack) • 2 Pressurized Mating Adapters attached to Unity
27 May 1999	2A.1	US Orbiter STS-96	• Spacehab—Logistics Flight
Nov 1999	1R	Russian Proton	• Zvezda Service Module
Dec 1999	2A.2	US Orbiter STS-101	• Spacehab—Logistics Flight
Feb 2000	3A	US Orbiter STS-92	• Integrated Truss Structure (ITS) Z1 • Pressurized Mating Adapter—3 • Ku-band Communications System • Control Moment Gyros (CMGs)
Mar 2000	2R	Russian Soyuz	• Soyuz • Expedition 1 Crew
Mar 2000	4A	US Orbiter STS-97	• Integrated Truss Structure P6 • Photovoltaic Module • Radiators
Apr 2000	5A	US Orbiter STS-98	• Destiny Laboratory Module
Jun 2000	5A.1	US Orbiter STS-102	• Logistics and Resupply; Lab Outfitting • Leonardo Multi-Purpose Logistics Module (MPLM) carries equipment racks
July 2000	6A	US Orbiter STS-100	• Rafaello Multi-Purpose Logistics Module (MPLM) (Lab outfitting) • Ultra High Frequency (UHF) antenna • Space Station Remote Manipulating System (SSRMS)
Aug 2000	7A	US Orbiter STS-104	• Joint Airlock • High Pressure Gas Assembly
Sep 2000	4R	Russian Soyuz	• Docking Compartment 1 (DC-1) • Strela Boom
Nov 2000	7A.1	US Orbiter STS-105	• Donatello Multi-Purpose Logistics Module (MPLM)
Jan 2001	UF-1	US Orbiter STS-106	• Multi-Purpose Logistics Module (MPLM) • Photovoltaic Module batteries • Spares Pallet (spares warehouse)

Table 4.1 (*Continued*) ISS assembly sequence: June 1999 planning reference.

Date	Flight	Launch vehicle	Element(s)
Mar 2001	8A	US Orbiter STS-108	• Central Truss Segment (ITS S0) • Mobile Transporter (MT)
May 2001	UF-2	US Orbiter STS-109	• Multi-Purpose Logistics Module (MPLM) with payload racks • Mobile Base System (MBS)
July 2001	9A	US Orbiter STS-111	• First right-side truss segment (ITS S1) with radiators • Crew & Equipment Translation Aid (CETA) Cart A
Aug 2001	11A	US Orbiter STS-112	• First left-side truss segment (ITS P1) • Crew & Equipment Translation Aid (CETA) Cart B
Nov 2001	9A.1	US Orbiter STS-114	• Russian provided Science Power Platform (SPP) with four solar arrays
Jan 2002	12A	US Orbiter STS-115	• Second left-side truss segment (ITS P3/P4) • Solar array and batteries
Mar 2002	12A.1	US Orbiter STS-117	• Third left-side truss segment (ITS P5) • Multi-Purpose Logistics Module (MPLM)
May 2002	13A	US Orbiter STS-118	• Second right-side truss segment (ITS S3/S4) • Solar array set and batteries (Photovoltaic Module)
June 2002	3R	Russian Proton	• Universal Docking Module (UDM)
July 2002	5R	Russian Soyuz	• Docking Compartment 2 (DC2)
July 2002	10A	US Orbiter STS-120	• US Node 2
Aug 2002	10A.1	US Orbiter STS-121	• Propulsion Module
Oct 2002	1J/A	US Orbiter STS-123	• Japanese Experiment Module Experiment Logistics Module (JEM ELM PS) • Science Power Platform (SPP) solar arrays with truss
Jan 2003	1J	US Orbiter STS-124	• Kibo Japanese Experiment Module (JEM) • Japanese Remote Manipulator System (JEM RMS)

Table 4.1 (*Continued*) ISS assembly sequence: June 1999 planning reference.

Date	Flight	Launch vehicle	Element(s)
Feb 2003	UF-3	US Orbiter STS-125	• Multi-Purpose Logistics Module (MPLM) • Express Pallet
May 2003	UF-4	US Orbiter STS-127	• Express Pallet • Spacelab Pallet carrying 'Canada Hand' (Special Purpose Dextrous Manipulator)
June 2003	2J/A	US Orbiter STS-128	• Japanese Experiment Module Exposed Facility (JEM EF) • Solar Array Batteries
July 2003	9R	Russian Proton	• Docking and Stowage Module (DSM)
Aug 2003	14A	US Orbiter STS-130	• Cupola • Science Power Platform (SPP) Solar Arrays • Zvezda Micrometeroid and Orbital Debris (MMOD) Shields
Sep 2003	UF-5	US Orbiter STS-131	• Multi-Purpose Logistics Module (MPLM) Express Pallet
Jan 2004	20A	US Orbiter STS-133	• US Node 3
Feb 2004	1E	US Orbiter STS-134	• European Laboratory—Columbus Attached Pressurized Module (APM)
Mar 2004	8R	Russian Soyuz	• Research Module 1
Mar 2004	17A	US Orbiter STS-135	• Multi-Purpose Logistics Module (MPLM) • Destiny racks
May 2004	18A	US Orbiter STS-136	• Crew Return Vehicle (CRV)
June 2004	19A	US Orbiter STS-137	• Multi-Purpose Logistics Module (MPLM)
July 2004	15A	US Orbiter STS-138	• Solar Arrays and Batteries (Photovoltaic Module S6)
Aug 2004	10R	Russian Soyuz	• Research Module 2
Aug 2004	UF-7	US Orbiter STS-139	• Centrifuge Accommodation Module (CAM)
Sep 2004	UF-6	US Orbiter STS-140	• Multi-Purpose Logistics Module (MPLM) • Batteries
Nov 2004	16A	US Orbiter STS-141	• Habitation Module

these flights there will be other automatic vehicle flights as well as Progress, Soyuz, etc, flights for crew transport, supply and logistics.

Notes

[1] This simulation of gravity will be used in large space stations, but it is different from what science fiction authors think of as artificial gravity, which would involve producing a gravity field without having the necessary mass.

[2] The craft did slowly spin in what was referred to as the *barbecue roll*. This was to ensure that each side of the spacecraft was exposed equally to the heat from the Sun and the cold of space. If the craft had kept the same attitude permanently, then that side facing the Sun would have baked.

[3] This is not so named because hitting the ground at terminal velocity would be fatal (although it undoubtedly would be). It is the velocity that is reached when acceleration stops as the forces have become balanced.

[4] The only situation that comes close occurred during the Apollo flights when, momentarily, the spacecraft passed through the point between the Earth and the Moon where the two forces of gravity exactly balanced. However, the Sun's gravity would still have been pulling on them even then.

[5] Douglas Adams in *The Hitchhiker's Guide to the Galaxy* refers to flying as 'learning how to throw yourself at the ground and miss'. This certainly applies to being in orbit.

[6] Of course, the required centripetal force also decreases as we make the radius of the circle bigger. However, the centripetal force decreases as $1/r$. Gravity goes down by $1/r^2$, so the gravity gets weaker more rapidly with distance than the centripetal force. To compensate the speed has to go down as well.

[7] Technically it is not really the boots that are providing the centripetal force. Recalling our discussion of the third law of motion in chapter 3, it is clear that the magnetic boots exert a force on the metal walls of the drum. It is the reaction force that the drum exerts on the boots, and hence on the astronaut, that forms her centripetal force.

[8] Strictly speaking it is the centre of gravity of the combination that is at one focus. However, for the Sun and the planets the Sun is so much more massive than all the planets put together that the centre of gravity of the solar system is virtually in the same place as that of the Sun.

[9] The tangent line at any point on a circle is a line drawn at 90° to the radius at that point on the circle.

[10] Of course, this motion is only genuinely horizontal, in the sense of being parallel to the ground, at apogee and perigee.

[11] The lunar equivalent of apogee.

[12] The lunar equivalent of perigee.

[13] This is an approximate average speed. The spacecraft would have started moving more rapidly and be slowed by the Earth's gravity. As it moved into the region where the Moon's gravity started to dominate, the craft would start to accelerate again.

[14] The same effect is no help on a Moon mission as the Moon is travelling round with the Earth.

[15] Aphelion and perihelion are the equivalents of apogee and perigee for orbits about the Sun.

[16] Ionizing radiations are emission that can ionize (remove electrons from) atoms. Some electromagnetic radiation—such as γ-rays—can do this, as well as the sort of particles produced by unstable atoms—radioactivity. A constant stream of such particles is produced by the Sun (called the solar wind). We are protected from the solar wind by Earth's atmosphere, although it is responsible for such effects as disturbing TV and radio communications and also the Aurora.

[17] The station, like any other object in orbit, lost energy through radiating it into space as infra-red electromagnetic radiation. Without the energy of sunlight warming it, this energy loss was not being replaced so the station cooled down.

[18] Some readers may remember that Kohoutek was billed as being a spectacular sight in the night sky—in actuality it could not be seen without a small telescope or binoculars.

Chapter 5

The Apollo command and service modules

The command and service modules were the heart of the Apollo spacecraft. The three astronauts had to live in the command module for much of the mission, supplied with air, water and electrical power from the service module. The service module's engine was responsible for slowing the craft into lunar orbit and boosting it away from the Moon on the return trajectory to Earth. The contract for the construction of the Apollo command module was awarded before the final decision was taken on which of the possible techniques for flying to the Moon was to be used. This made significant progress on the design difficult as the demands on the capsule varied depending on the technique used. In the end the problems of designing a multi-purpose capsule (which they could make good at many things but not excellent at anything) helped to push the mission planners into making a decision. The first part of this chapter outlines the debate that took place within NASA regarding the fundamental questions of mission design.

5.1 Mission modes

When President Kennedy issued his challenge in May 1961, NASA's long term strategy planners had already been considering ways of landing a man on the Moon. On the grounds of minimizing risk, many of them had decided on a direct flight that did not involve docking spacecraft. Such a mission would launch a huge rocket, the upper stage of which would fly directly to the Moon. Once there, it might go into orbit temporarily then land, or it might proceed directly to the surface. Having completed their exploration the astronauts would then fire the engines again and use the whole craft to return to Earth.

Docking manoeuvres were thought to be needlessly risky as they required delicately flying two spacecraft in close proximity. At the time NASA had no experience of flying spacecraft with the precision necessary for docking. The Mercury astronauts had shown that control was possible in space, but this would be a whole new degree of flying[1]. Also there was the complexity of finding the other craft in the first place! However, direct flight had its drawbacks as well. As there was no plan to separate and discard parts of the craft that would serve no purpose on the return journey, considerable demands would be placed on the landing legs. If the whole craft were to return then enough propellant to do the job would need to be carried, making the craft extremely heavy. The legs would have to survive the shock of landing such a heavy craft. The engines would also have to be powerful as they would be needed to land the heavy vehicle softly and lift it back to Earth again. On top of all this the booster rocket needed to launch the whole mission from the surface of the Earth would be much bigger than anything that NASA had developed so far.

At the time the only significant competition to direct flight came from a group that used to work outside NASA—Von Braun's ex-army missile group. His team favoured an *Earth Orbit Rendezvous* (EOR) mission. Von Braun's vision was to fly several boosters into Earth orbit where the Moon ship could be constructed and fuelled. From this point on the mission would fly in exactly the same manner as the direct flight. The advantage was that the fully fuelled ship would not have to be lifted from the ground in one go. After his group transferred from the army to NASA in 1960, effectively becoming the Marshall Space Flight Center, they continued to work on the EOR mission profile.

The Marshall Space Flight Center was not the only NASA facility with ideas on how to get to the Moon. From about 1959 the Jet Propulsion Laboratory (JPL) in Pasadena, California, was championing *Lunar Surface Rendezvous*. This idea involved sending a number of unmanned rockets to the Moon in advance of the astronauts. One would be an Earth return craft and it would be accompanied by one or more fuel tankers. This would enable the ground control team to refuel the Earth return vehicle automatically and confirm that it was in a fit state for the return flight before sending the astronauts out. This plan was considered too risky by senior management at NASA, but it is interesting to note that a similar idea is now being seriously considered for a future Mars mission (see chapter 8).

Exercising the minds of all the mission planners was America's lack of a large payload booster in the early 1960s. The first American in space had been launched by the Mercury Redstone (a modified ballistic missile). The Atlas and Titan II boosters developed for later Mercury and Gemini missions were not powerful enough for Apollo. Plans for two new boosters were underway, the *Saturn* and the even bigger *Nova* vehicle. Saturn was envisaged as a family

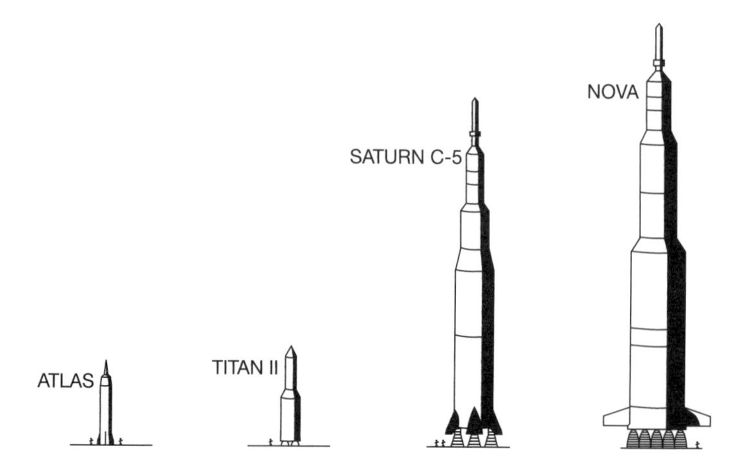

Figure 5.1 The proposed development of American rockets in the early 1960s.

of boosters that would be capable of launching Earth orbit and lunar orbital missions. Von Braun's EOR flight plan would involve launching several Saturns (von Braun felt that developing Nova would be pushing the technology too far in the time allowed). Nova was developing in parallel as a booster capable of hurling a spacecraft to the Moon on a direct flight mission. This vast booster would use eight F-1 engines in its first stage. However, when Kennedy laid his challenge on the world stage the F-1 engine and the Saturn were in need of a great deal of development. Decisions about how the booster programme should continue were affected by the mission profile discussion, which was in turn influenced by the booster plans. The circle needed to be broken.

With the pressure on to land a man on the Moon as soon as possible, some quite wild ideas were being touted about. Everything was considered, from refuelling the spacecraft *en route* to the Moon to landing a man on the surface and keeping him alive with regular supply rockets until NASA could think of a way of getting him home!

One idea sparked in this period was destined to be rather more significant—that of *Lunar Orbit Rendezvous* (LOR). In this scheme a lunar lander would detach from the spacecraft carrying astronauts to the surface. When the time came to leave, the lander would take off from the surface and dock with the 'mother ship' which had remained in orbit. The crew could then transfer to the command module for the return to Earth. When this idea was first mooted in early 1959 it was seen simply as a way of reducing the total weight of the spacecraft, as components could be abandoned at appropriate moments in the mission. This idea carried the staging of the booster to its logical conclusion. However, against it was the fact that one of the most mission-critical manoeuvres, docking with the command module, had to take place in lunar orbit—a very long way from mission control and with very few back-up options to save the crew should anything go wrong.

In the NASA history of the development of the Apollo spacecraft (*Chariots for Apollo*, see the reference in the bibliography at the end of this book), Thomas Dolan of the Vought Astronautics Division is credited with the first design study of the LOR mission mode. He assembled a team to look into lunar missions and ways in which his company might contribute to NASA's plans. As it turned out, Vought was not awarded any of the contracts for Apollo, but their comments turned out to be prophetic. Dolan's team fully appreciated that ΔV budgets are the key to mission design. They proposed a modular spacecraft, with separate components performing different functions. Dolan concluded that the best approach was to discard the pieces that were no longer needed. Furthermore, he saw no reason to take the entire spacecraft down to the lunar surface and back to lunar escape velocity. All this was done in 1959.

Between 1959 and 1960 another NASA agency, the Langley Research Center, formed several committees to look into the role of rendezvous as part of normal operations for a space station in Earth orbit. One of these committees was headed by John Houbolt. His group wanted to broaden their remit to look into possible Moon landing missions. A more formally established lunar mission steering group was convened at Langley during 1960. One of its members, William Michael, produced a monograph that described the advantages of parking an Earth return propulsion system in lunar orbit while a low-mass lander took members of the crew to and from the surface. This report ended up on Houbolt's desk and he became an instant convert to the virtues of LOR. Langley's official position turned to LOR as the means of landing men on the Moon. Through 1960 and 1961 Houbolt did the rounds of various NASA meetings and committees trying to convert people to the LOR scheme. However, he met with little success. Finally in May 1961 he wrote directly to the NASA operations vice president (effective second in command) Robert Seamans. Initially this had little effect. As July and August wore on, Houbolt and the Langley teams began to get discouraged at the lack of impact that their ideas were apparently making in NASA. There was a brief moment of optimism when he was asked to prepare a paper on rendezvous for the Apollo Technical Conference in mid-July of that year. Unfortunately when the time came for him to submit his paper for final acceptance he was told to concentrate on the general principles of rendezvous techniques and to 'throw out all that LOR'.

The break-through finally started to be made when the Golovin Committee was set up to look into the booster development program. Nicholas Golovin's team were supposed to recommend a set of booster designs for development, but discovered that such decisions were inextricably linked to questions of how the boosters were to be used—in other words the manner in which the mission to the Moon was to take place. On 29 August 1961 Houbolt and two more Langley employees presented their ideas to the Golovin committee. Buoyed by the interested response of the committee members and LOR's favourable mention in the committee's final report, Houlbolt prepared a report for submission directly to Seamans. The opening phrases in that report have become somewhat famous after their inclusion in the HBO series *From the Earth to the Moon*:

> 'Somewhat as a voice in the wilderness, I would like to pass on a few thoughts on matters that have been of deep concern to me over the recent months.'

This time Houbolt got an encouraging reply:

> 'I agree that you touched upon facets of the technical approach to manned lunar landing which deserve serious consideration ... It would be extremely harmful to our organization and to the country if our qualified staff were unduly limited by restrictive guidelines.'

LOR had moved onto the table and from that point started to be given equal consideration to the other schemes being considered.

However, it still took some while for NASA to finally decide that LOR was the most sensible way of getting men to the Moon within the time constraints imposed by the President. There was a great deal of opposition within NASA to this plan, stemming from concern about the safety of the crew. Not only were there the problems of finding and docking with a spacecraft in orbit round the Moon (it was seen as being bad enough in Earth orbit), but also there was no obvious abort mechanism. In the EOR mode, if the ship could not be fuelled, then the crew simply returned to Earth.

One of the arguments that gradually swung support towards LOR was the problem of designing a ship that could carry the crew to the Moon, land safely on the surface and return to Earth. As designers worked on the problems of manoeuvring a craft near to the surface, other issues, aside from the weight of the craft, began to emerge (such as being able to see the lunar surface clearly as the craft was landing!). It became clear that combining the roles of crew transport and lander forced design compromises on the craft that made it difficult to achieve either role satisfactorily.

The key moment in NASA's conversion to LOR came when Von Braun turned to the idea after having being one of the most vigorous proponents of EOR. He announced his conversion in June 1962 at a meeting convened to present the results of the Marshall Space Flight Center's continuing studies into EOR. After having listened to various members of his team describe their EOR plans, von Braun stood up and spoke from notes made during the meeting. To his astonished audience he announced that the centre would support LOR. He explained his conversion in the following way:

> 'I would like to reiterate once more that *it is absolutely mandatory that we arrive at a definite mode decision within the next few weeks*... If we do not make a clear-cut decision on the mode very soon, our chances of accomplishing the first lunar expedition in this decade will fade away rapidly.'

He continued to say that LOR 'offers the highest confidence factor of successful accomplishment within this decade'. The key reason for this was the separation between the design of the landing craft and the command module: 'A drastic separation of these two functions into two separate elements is bound to greatly simplify the development of the spacecraft system [and] result in a very substantial saving of time'.

From this point on all the NASA centres converged on LOR as the way to get a man to the Moon.

5.2 The command module

The contract for the design and manufacture of the Apollo command module was granted before the decision to go with LOR had been taken. By late 1962 plans were well underway for the command module of an Earth-orbital and circumlunar craft. The contractor, North American, had given some thought to attaching a propulsion stage for a landing on the Moon. However, this need was circumvented by the switch to LOR.

The greatest effect that the decision to go with LOR had on the design of the command module was the need to incorporate a docking mechanism with the ability to transfer two astronauts to the lunar module (LM). Consequently development of the command module continued along two parallel courses. The early design, not including docking facilities, became the Block I spacecraft that was used in astronaut training and was destined for Earth-orbital missions. However, due to the pressure of time imposed by the Presidential deadline, North American could not wait for the Block I ship to be tested and flown before work on the Block II (docking version) could begin. In fact the Block I design never flew a manned mission[2].

Clearly some components were common to the two versions, and North American sub-contracted for systems such as the escape tower with its solid rockets, the command module manoeuvring thrusters, and the heat shield.

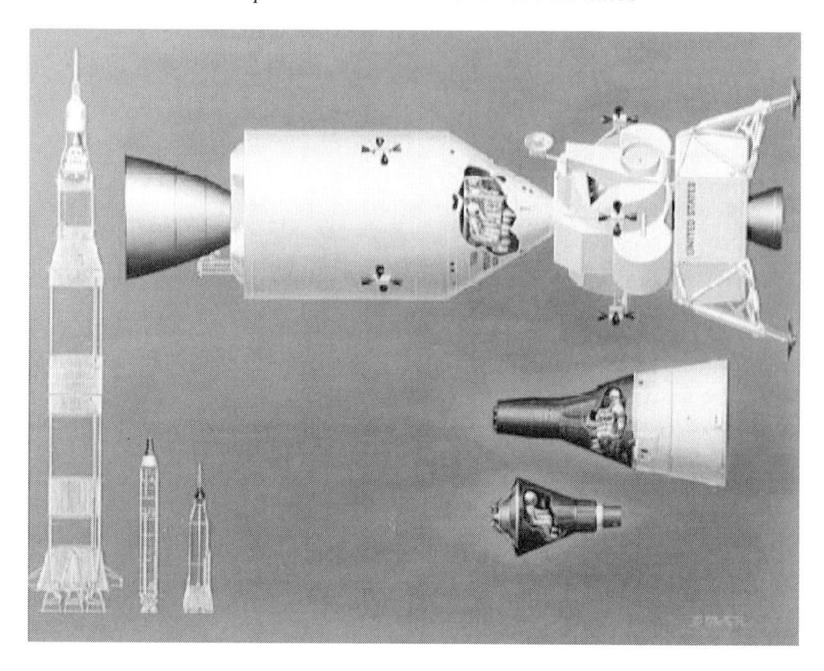

Figure 5.2 In this NASA graphic the huge advance in spacecraft design implied by Apollo is illustrated. On the left the Saturn, Titan II and Atlas boosters are compared. The jump in size between Mercury, Gemini and Apollo is illustrated on the right.

The basic design parametres for the command module called for it to function as a 'combination cockpit, office, laboratory, radio station, kitchen, bedroom, bathroom, and den'. The specification also called for a 'shirtsleeve' environment within the capsule as it was unrealistic to expect the astronauts to remain in their space suits for the duration of the journey. Designers rapidly settled on the Apollo command module being a refinement of the Mercury capsule used at the start of the manned space programme (see figure 5.2).

The blunt-ended cone shape of Mercury had been shown to be effective in surviving the rigours of re-entry. Calculations showed that the Apollo capsule should be more tapered than Mercury had been. The higher entry speed experienced by a ship returning from the Moon would generate a shock wave in the atmosphere. This in turn would cause a super-hot region to form around the capsule—so hot that the heating effect of the infra-red radiation produced would significantly raise the external temperature of the craft. This extra heating mechanism (in addition to conduction and convection in the atmosphere) was not experienced by the slower-entering Mercury capsules. The more tapered shape helped to mitigate this effect.

Figure 5.3 Testing the launch escape system which was designed to pull the whole command module clear of the Saturn V should a problem develop during launch. The command module would then use its parachutes to land.

The choice of splashdown into the ocean rather than a land-based return was forced by the geographic position of the launch site. Kennedy Space Center being on a peninsula (for safety reasons—large potentially dangerous bombs (rockets) are not launched near major population centres) meant that if an abort took place during the launch the capsule would have to land in the water anyway, so why not make that the preferred landing option?

Command module construction and heat shields

Dimensions:
Height: 3.22 m
Diameter: 3.91 m
Mass (including crew): 5900 kg
Mass (splashdown): 5310 kg

Once assembled the command module sub-divided into three compartments. The forward compartment was the small space at the point of the cone. The crew compartment was within the primary structure at the centre of the capsule, and the aft compartment was another small space that ran around the outside edge of the lower half of the primary structure.

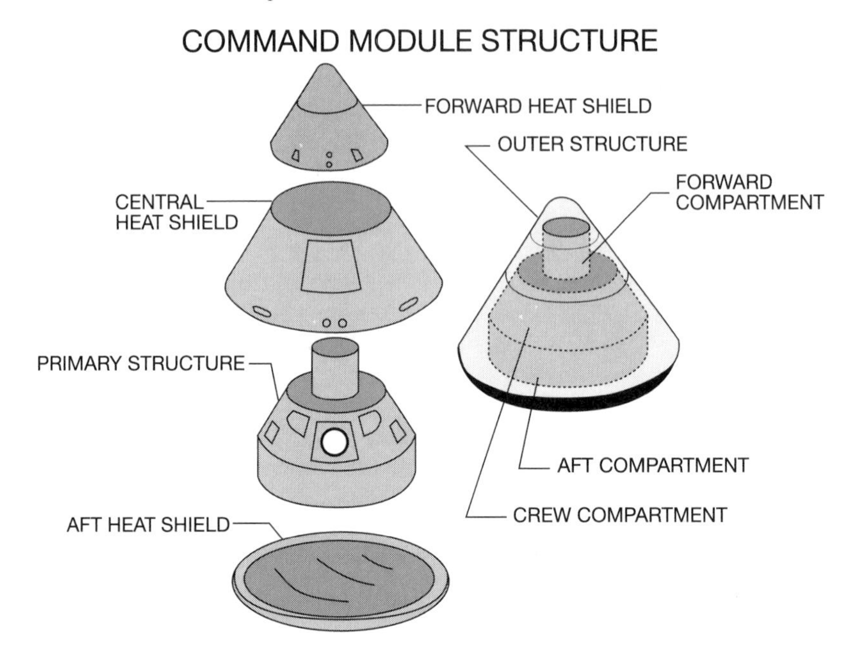

Figure 5.4 Design of the Apollo command module.

The primary structure housed the astronauts and had to be a pressurized 'shirtsleeve' environment. It was constructed from a welded aluminium inner skin to which was glued an aluminium honeycomb[3] faced with further sheet aluminium. The honeycomb varied from being 6.4 cm thick at the top (near the docking structure) to 3.8 cm thick near the aft heat shield. At the top of the primary structure was a cylinder that reached to the very top of the capsule. This was the tunnel through which the mission commander and lunar module pilot would crawl into the lunar module while it was docked to the command module. Prior to docking the probe mechanism was located at the end of this tunnel.

Forming the sloping sides of the cone was the outer structure, manufactured from stainless steel brazed into a honeycomb and in turn brazed to the steel alloy inner and outer sheets. The outer structure formed one of the principal barriers to protect the astronauts and their equipment from the extreme temperature variations that would be present during the mission. Consequently a heat shield was bonded to its outer surface.

During lift off, friction with the atmosphere raised the external temperature to 650 °C. Once in space, the side of the spacecraft facing towards the Sun would be baked by sunlight to 140 °C, and the side away from the Sun would freeze to −170 °C. Finally, during re-entry the external temperature would rise to 2800 °C—although the brunt of this would be taken by the aft heat shield.

In flight the spacecraft was set into a slow rotation (barbecue roll) which ensured that each side received an even baking from the Sun.

The heat of the launch was dealt with by the boost protective cover. This was another cone, made from fibreglass with a white-painted cork outer cover, that fitted over the command module and was fixed to the launch escape tower. Fortunately, the launch escape system never had to be used during a mission—it was ejected along with the boost cover at about 90 km.

Bonded onto the outer structure was the heat shield which dealt with re-entry. This comprised 400 000 plastic honeycomb cells which were filled, by hand, with an ablative material designed to protect the crew from extreme temperatures. Such materials char and melt, rapidly absorbing and dissipating energy as they do so. At the time there was no material that could fully prevent thermal conduction through to the astronauts at such high temperatures[4]. By progressively melting and falling off, an ablative heat shield can form a very effective protective barrier.

The aft heat shield was constructed in the same manner as that over the sides of the cone; however it was somewhat thicker (up to 5 cm). The final outer covering over all the heat shields was a moisture barrier and a Mylar thermal coating that looked like aluminium foil.

The completed command module was as light as possible, yet rugged enough to survive the heat of re-entry, the impact of splashdown and possibly collisions with small meteorites.

The forward compartment

This was the section of the command module that ran in a ring round the docking tunnel at the top of the primary structure. It was divided into four equally sized segments containing all the equipment necessary for landing (parachutes, beacon light, recovery antenna, floatation balloons, etc), two small manoeuvring motors and the release mechanisms for the forward heat shield (this had to be ejected before the parachutes could come out). In turn the forward heat shield contained four slots into which the legs of the launch escape tower would fit. The tower was attached with bolts that contained small explosive charges. When the time came to jettison the launch escape tower and the boost cover the charges would fire, breaking the bolts.

Figure 5.5 Stages in the construction of a command module. Top, the cabin section (or primary structure) of the CM is assembled at North American in 1965; bottom, a completed central heat shield is lowered into place over the primary structure in May 1966.

The aft compartment

Running round the outside of the lower part of the primary structure and above the aft heat shield, the aft compartment was divided into 24 bays. Enclosed in these bays were 10 more manoeuvring engines with the fuel and oxidizer tanks for the whole command module manoeuvring system, water tanks, some ribs that were designed to crush on impact with the water (just like the deforming structure in a car which absorbs energy in a crash), some instruments and the point where wiring and plumbing ran from the command module to the service module.

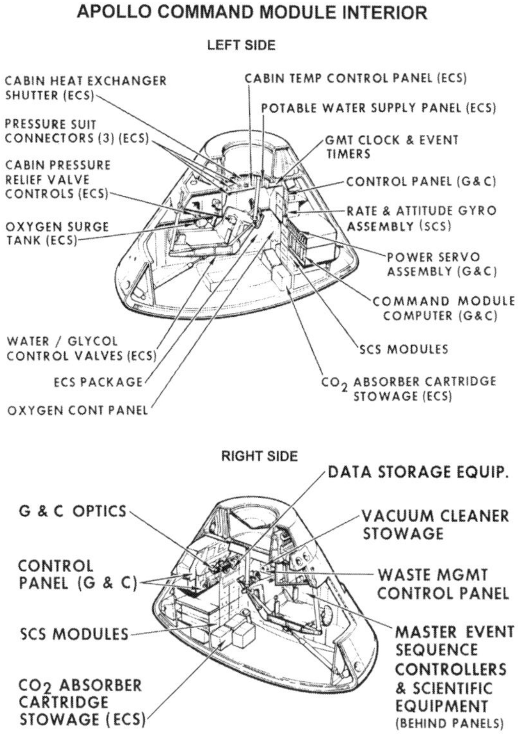

Figure 5.6 The interior of the command module (the centre couch is not shown).

Inside the command module

The crew compartment inside the primary structure contained the living and working environment for three men during the majority of the mission. Indeed, the command module pilot would never normally leave the crew compartment during the whole mission as he was left in orbit while his two colleagues took the LM down to the surface. Inside was a habitable volume of just under six cubic metres.

The requirement that the command module be a home to three astronauts for several days placed new demands on the design of atmospheric conditioning equipment. The cabin was air conditioned with the temperature regulated to between 21 and 24 °C. A pure oxygen atmosphere was used with a pressure of 35 kPa in flight[5]. Exhaled gases were recycled and the CO_2 contaminants removed so that the remaining oxygen content could be put back into the atmosphere[6]. The unit also provided cooling water that circulated through the

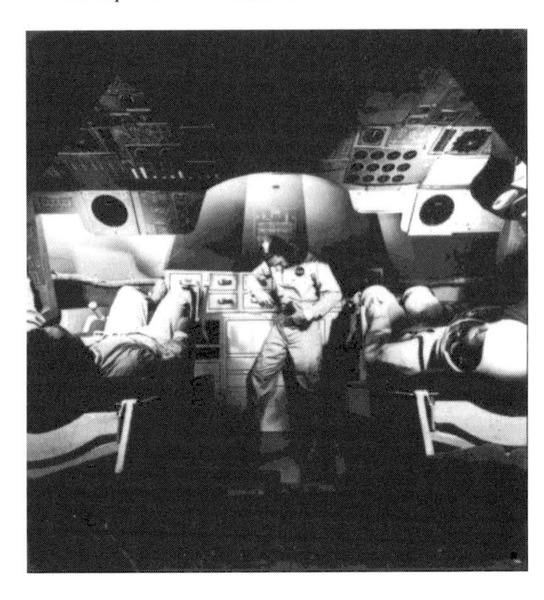

Figure 5.7 An early mock-up of the inside of the command module showing the centre couch removed. The central control panel can be seen running across the top of the picture.

electronic equipment in the command module, preventing it overheating due to its own operation. On Earth the equipment could be expected to keep itself quite cool by convection of the surrounding atmosphere. Energy conducted into a gas causes its molecules to speed up, which in turn lowers the density of the gas as the molecules spend more time further away from each other. The portion of the atmosphere nearest to the equipment is obviously heated first, so it becomes a lower density than the portions next to it. In a gravitational field this causes the denser air to sink and buoy up the less dense gas. This is how a convection current works. However, in a spacecraft on the way to the Moon, gravity is not playing much of a part so the convection currents are very sluggish[7]—hence the need for a circulating coolant that can conduct excess energy away. The unit that achieved all this was a masterpiece of design. Redundant parts were extensively used to ensure reliability, yet the whole unit was little bigger than a large microwave oven. Excess heat extracted from the command module was radiated into space by two large radiators mounted on the service module.

Across the centre of the crew compartment were the three couches on which the astronauts would spend much of their time. They were designed to support them during the stress of launch and splashdown and were connected to the command module's structure by deformable struts (same idea as the ribs in the aft compartment and the springs illustrated in section 2.5). Looking in from the entry hatch in the sloping side of the capsule, the left most couch was occupied by the mission commander. In the centre was the command module pilot, who was responsible for navigating and manoeuvring the spacecraft during some mission-critical moments (such as docking with the LM and extracting it from the Saturn V third stage). The lunar module pilot occupied the right-hand couch. His main duty was to manage the various sub-systems in the spacecraft.

The seat portion of the centre couch could be folded, allowing more room to move about. Sleeping bags were slung under each of the outer two couches.

Running across the command module in front of the couches was the main instrument and control panel. This was just over two metres long and a little under a metre deep with two 'wings' at either end which were 90×60 cm. The panel was divided into sections with specific controls and dials placed in front of the couches where the crew member responsible would normally be placed.

The flight controls were positioned before the mission commander. He could see instruments that informed him about the spacecraft's altitude, velocity and attitude as well as the controls for propulsion, crew safety, splashdown and emergency detection of the capsule (in case the splashdown was some way off target). One of the two navigation and guidance computer panels was mounted in this section of the panel as well. Controls for manoeuvring the spacecraft (including a small joystick) were mounted on the arms of the mission commander's and the command module pilot's couches.

The command module pilot faced the centre of the panel. He could reach some of the controls mounted on his colleague's stations to either side of him as well as his own specific instruments and controls directly in front of him. The latter included controls for managing the propellant for the command module manoeuvring engines, environmental controls, controls and instruments relating to the cryogenic storage of liquid oxygen and hydrogen in the service module[8], as well as various general warning systems.

Finally, the lunar module pilot's area contained communications, controls and instruments related to the electrical power systems and fuel cells, data storage and propellant management for the service module engine.

Great care was taken in the design of the control panel and the instruments mounted on it. Many modifications were made as a result of astronauts' suggestions made during mission simulations. A great deal of training was done inside command module mock-ups (and of course inside similar simulators for the lunar module) in which the crew would rehearse every aspect of their flight. Often problems with the design of the panel would be found during such simulations. All the controls were designed to be operated by the crew wearing

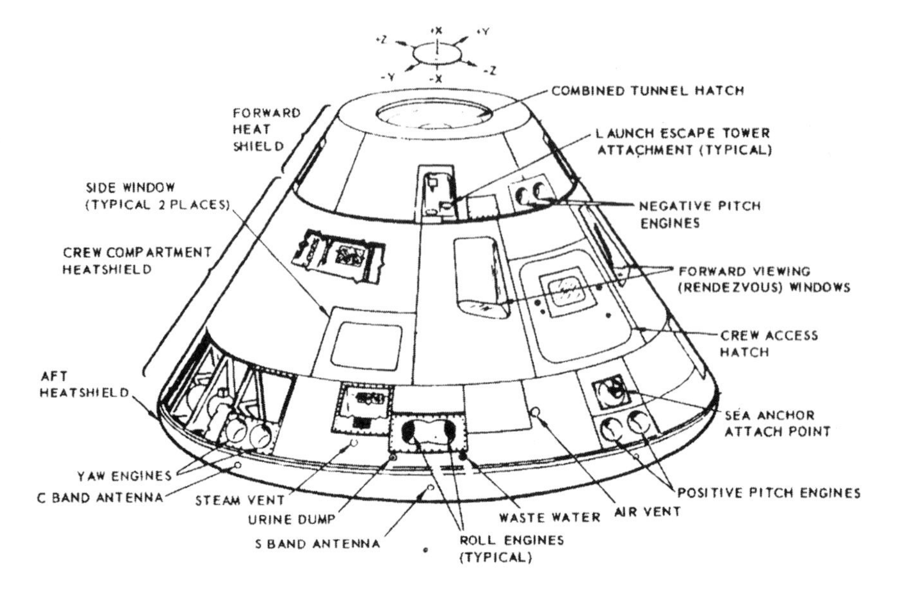

Figure 5.8 An exterior view of the command module showing some of the reaction control thrusters.

their spacesuit gloves. The most critical switches and controls were guarded with flip-up covers to prevent them being activated by accident; some even had locks that had to be released before they could be activated.

Manoeuvring

While in space the combined command and service module units could be manoeuvred about by two systems. Large course corrections, such as capture into lunar orbit, were achieved by firing the main service propulsion system (SPS). Smaller changes, docking manoeuvres and setting up the barbecue roll were done with the reaction control system (RCS) on the service module. However, once the service module had been jettisoned (about 15 minutes before entry into Earth's atmosphere) the reaction control systems on board the command module had to take over to correct the path through the upper atmosphere.

The command module was equipped with two independent systems of six thrusters (one acting as a standby in case the primary system failed) dotted round the module in the aft and forward compartments. They used a hydrazine fuel/nitrogen tetroxide oxidizer combination (which is hypergolic) and each engine provided 400 N of thrust.

Docking

As the main argument used against the LOR and EOR mission modes had been the difficulty of docking two spacecraft in orbit, it is hardly surprising that a great deal of careful design went into the system used to link the command and lunar modules.

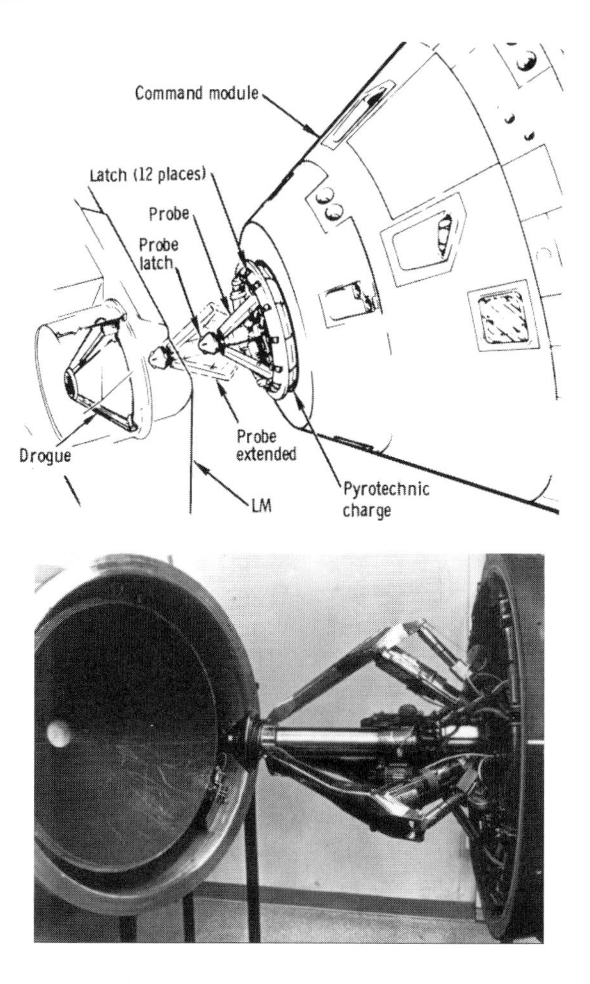

Figure 5.9 The docking mechanism.

The system consisted of a probe mounted on the end of the command module and a dish-shaped receptacle (drogue) on the lunar module. The idea was to manoeuvre the two modules together so that the probe entered the drogue. The probe then slid down the inside slope of the drogue until the latches at the end engaged with the socket at the base of the drogue. When that happened a partial dock had been achieved. The astronauts then operated the probe retraction system that pulled the probe back in towards the command module, bringing the docked lunar module with it. At contact between the two, 12 automatic latches in a ring round the outside of the probe mechanism engaged, forming an airtight seal between the two spacecraft. This was known as a *hard dock*.

Immediately behind the probe on the CM side and the drogue on the LM side were two halves of a tunnel that enabled astronauts to transfer between the two spacecraft. The halves were terminated by hatches on either side. Before the command module hatch could be opened, the pressure difference between the interior of the capsule and the tunnel had to be eliminated by operating a valve mechanism.

Next, the CM hatch was removed and one of the astronauts checked the 12 latches, manually activating any that had not tripped. Next he connected an electrical cable between the CM and LM. He then floated down the tunnel until he reached the reverse side of the docking probe mechanism at the 'top' of the command module tunnel. This could be removed with the drogue still connected to the other side (remember that it is now the 12 latches holding the spacecraft together!) and passed back to be stowed on one of the couches in the command module. Finally he operated another valve to equalize the pressures in the tunnel and the LM before opening the last hatch. This one was not removable—it hinged inwards into the LM.

The sequence was slightly different when the CM and LM docked for the second time, in lunar orbit after the LM ascent stage had lifted off from the Moon. This time, once hard dock had been achieved, the command module pilot opened the CM hatch and removed the probe at his end while his colleagues in the lunar module opened their hatch and removed the drogue from inside the LM.

When it became time to separate the two vehicles for the landing the two astronauts inside the LM were passed the drogue by the command module pilot who then replaced the probe and disconnected the cables. Finally, he manually cocked all 12 latches (not releasing them, which could have been embarrassing) and closed the hatch. Separation was achieved by electrically releasing the latches from inside the LM.

When the LM was finally jettisoned after the Moon walkers had returned to orbit, release was achieved by explosively blowing away the whole CM docking ring (on which the latches were mounted) with small pyrotechnic charges. This in turn pushed the LM away from the CM.

The first docking that took place during the mission was the extraction of the LM from its housing in the third stage of the Saturn V. This took place

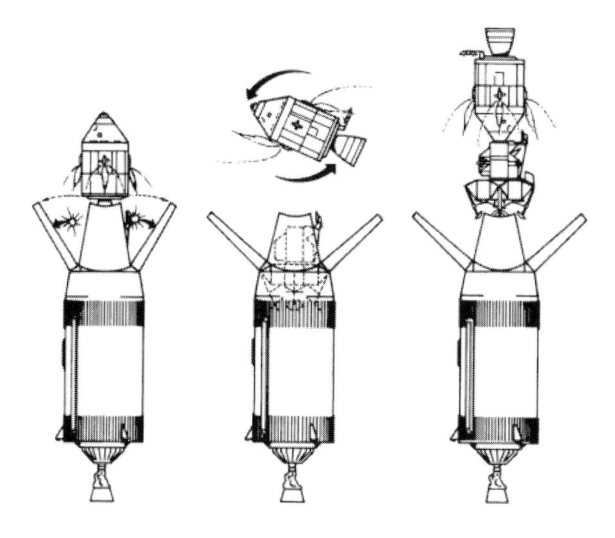

Figure 5.10 The sequence leading up to lunar module extraction.

about 25 minutes after the main engine burn to inject the spacecraft into its lunar trajectory. Explosive charges separated the command and service module combination (CSM) from the third stage and then the LM covers were blown open. The command module pilot used the thrusters on the service module to push the CSM ahead of the stage, rotate it and then come back to dock with the LM.

Optical aids ensured that the command module pilot had a means of accurately judging the speed of approach and alignment with the target on the LM. (During docking in lunar orbit the command module pilot had to watch while the manoeuvring was done by the LM crew.) Once a dock had been made, springs that joined the LM's legs to the third stage were released and these helped to push the docked spacecraft away from the stage. Thrusters were then used to pull the combination further ahead.

The Apollo 14 mission was nearly brought to an end soon after it started when problems developed during docking. The command module pilot Stu Roosa had trained hard for this manoeuvre and badly wanted the record for the smallest amount of propellant used. He manoeuvred the CSM into the LM drogue perfectly first time, but it refused to make the initial latch. Three more attempts were made with no more success. Although in principle the CSM could coast along in front of the third stage indefinitely, the third stage was programmed to vent any remaining propellant into space after a certain time and it would be far safer for the Apollo spacecraft to be well out of the way when that happened. On board the command module the mission commander, Alan Shepard, was considering options such as depressurizing the CM to retrieve the probe and attempt to fix it. As a final option he was prepared to pull the spacecraft together by hand. The problem lay in the initial latch. He was sure if he could just get the craft to join, the 12 main latches would trigger.

In the end mission control suggested that they have one more attempt, but this time not to gently bring the craft together. They suggested that the impact be harder, which would automatically contract the probe mechanism bringing the latches into play. We will never know what the determined Alan Shepard might have done, for the final 'charge' attempt secured a hard dock and the mission proceeded.

Re-entry and splash down

The command module entered the Earth's atmosphere at about 11 km s^{-1} and at an angle that was critical to the crew's survival. The drag on the capsule passing through the atmosphere at such enormous speeds decelerated it at up to $6g$. At the same time the CM's aerodynamic qualities produced lift in a similar manner to an aeroplane wing generating a net upward force[9]. The balance between drag and lift varies depending on the angle of flight.

If the capsule entered the atmosphere at too small an angle then the lift would climb quickly with little drag to slow it down—essentially because the spacecraft was not entering dense atmosphere quickly enough. This would cause it to skip off the atmosphere into space—rather in the manner of a stone skipping off water. If the capsule entered at too great an angle, the drag force would increase rapidly, slowing the spacecraft down. As lift is dependent on speed, the capsule would dip into the dense atmosphere quickly. The drag would increase and there would be a danger of the heat shield failing. The spacecraft had to hit the atmosphere at an angle of at least 5.2° and not more than 7.2°: an 'entry corridor' that was only 2° wide (see figure 5.11).

The re-entry path was flown automatically by the on-board computer using the command module's thrusters to steer. As figure 5.11 shows, the path was designed to drop into the dense air, then rise up again briefly to give the crew and spacecraft a brief rest from the extreme temperature and g force (rising to $6g$) before diving in again to complete the deceleration. If there was a failure in the guidance system, the astronauts were trained to fly re-entry themselves (as Shepard had done on the first manned orbital flight—see intermission 6). Of course, no human could fly the craft as accurately as the computer, so the chances were that they would splash down some way from the target point, but at least they would be alive.

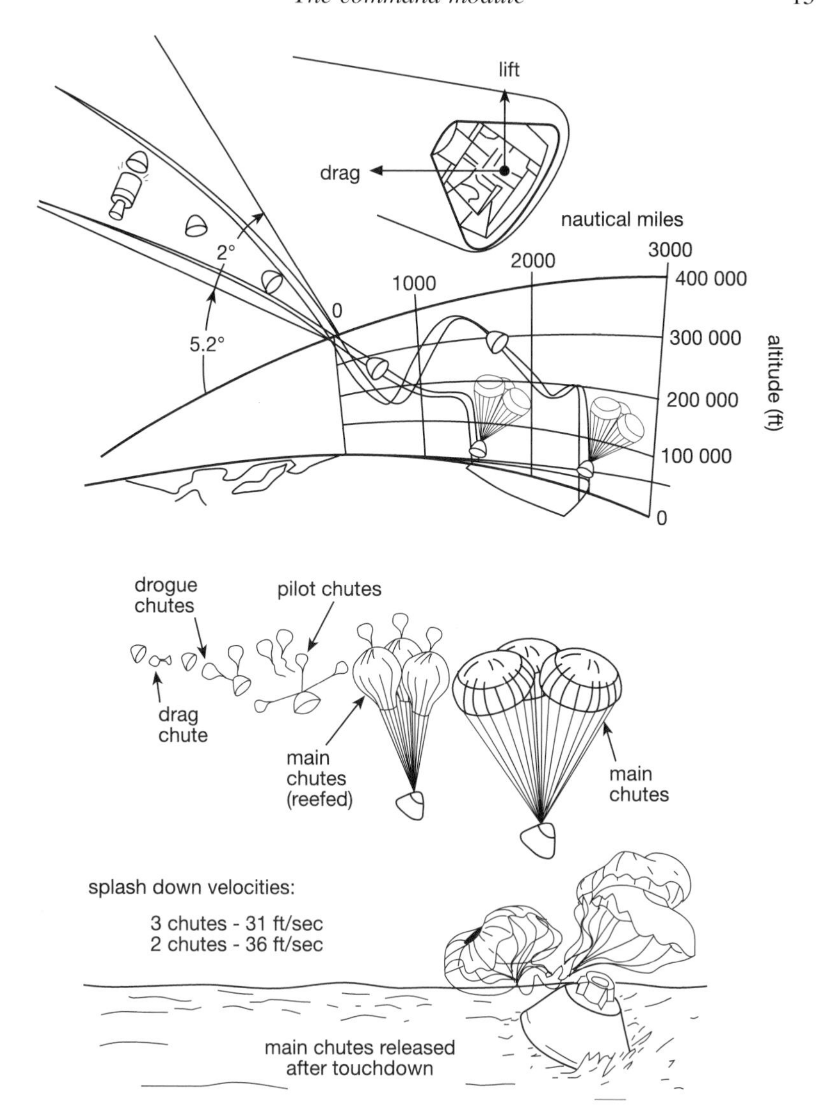

Figure 5.11 The entry corridor and the sequence of events after aerodynamic deceleration.

During re-entry the Apollo command module formed the head of a trail of super-hot gas 200 km long.

Having emerged from the enormous temperatures of re-entry, the command module would have slowed to a more sedate 480 km h^{-1}. The next stage of deceleration was achieved by parachute. First the forward heat shield was ejected from the top of the command module (at about 7 km altitude) to allow the drogue 'chutes to be deployed 1.5 seconds later. These were 5 m in diameter and slowed the capsule to 280 km h^{-1}. Their other job was to ensure that the capsule was oriented at the correct angle for the remaining descent. Once the altitude had dropped to 3 km the drogues were released and the pilot 'chutes took over. Their role was to pull the main three-parachute system from the command module. The main 'chutes were enormous—25 m in diameter—and the capsule hung 36 m beneath them. By the time the command module had reached the water the main 'chutes had slowed it to 34 km h^{-1}.

The great tragedy

On 27 January 1967 a tragedy struck the preparations for Apollo that was unprecedented in the history of the manned space flight programme. Three astronauts, Gus Grissom, Roger Chaffee and Ed White were in full space suits strapped into their seats on board a Block I command module atop a Saturn Ib booster on pad 34. They were there for a countdown dry run. It was to be a complete test of the systems both on board the spacecraft and in the control rooms prior to the scheduled first manned mission for Apollo—slated for February. The test had been long and arduous. The command module had shown several bugs in its components and a maddening fault with the communications systems. However, for one brief, terrible moment contact with the control rooms was clear and unambiguous. At 6.31 pm, about five and one half hours into the test, a short transmission was sent from the command module that froze the blood of all who heard it: 'Fire'.

Within moments this first hint of trouble was followed by 'We've got a fire in the cockpit' and then, with mounting urgency 'We've got a bad fire... We're burning up'. Television monitors trained from the outside at the command module hatch showed the crew struggling to release the mechanism. On the pictures a bright glow inside the capsule and flames licking out from round the hatch could be seen through billowing smoke. Eventually the hatch window was completely obscured by the smoke inside. Meanwhile five people outside the spacecraft were fighting the intense smoke, heat and flames to reach the crew and also struggled with the hatch mechanism. They finally gained access to the inside of the command module five minutes after the first indications of trouble. By that time it was too late. All three astronauts, scheduled to be the first men to fly Apollo into Earth orbit, had died on the launch pad.

Post mortems revealed that the space suits had protected the crew from most of the fire. They had second- and third-degree burns, but they were not in themselves fatal. What killed them was asphyxiation. The suits were connected to the command module's systems by breathing hoses that were passing a pure oxygen atmosphere through to the crew. When the fire burned through these hoses, the smoke and other fumes passed directly into the spacesuits. The astronauts at that point died in moments.

In the weeks and months that followed 27 January 1967, a day that profoundly affected the lives of many within NASA, the strenuous investigations into the incident not only revealed the likely cause of the fire, but also brought before management's eyes, in a forceful manner, complaints that the astronauts had been making for some time.

The NASA internal review board issued a summary report[10] on 5 April 1967 in which they said that:

> *'Although the Board was not able to determine conclusively the specific initiator of the Apollo 204 fire, it has identified the conditions which led to the disaster:*
>
> - *A sealed cabin pressurized with an oxygen atmosphere.*
> - *An extensive distribution of combustible materials in the cabin.*
> - *Vulnerable wiring carrying spacecraft power.*
> - *Vulnerable plumbing carrying a combustible and corrosive coolant.*
> - *Inadequate provisions for the crew to escape.*
> - *Inadequate provisions for rescue or medical assistance.*
>
> *Having identified the conditions that led to the disaster, the Board addressed itself to the question of how these conditions came to exist. Careful consideration of this question leads the Board to the conclusion that in its devotion to the many difficult problems of space travel, the Apollo team failed to give adequate attention to certain mundane but equally vital questions of crew safety. The Board's investigation revealed many deficiencies in design and engineering, manufacture and quality control.'*

It seems that no-one had seriously considered the hazards of an over-pressurized cabin full of pure oxygen. A pure oxygen atmosphere was used as manipulating two or more gases would have involved heavy and complex additions to the cabin environment systems. Fire risks in flight had been studied extensively, but no-one had thought much about the situation on the ground. While on the pad, the pure oxygen environment at greater than atmospheric pressure had been chosen to flush any contaminants out of the command module.

In an oxygen atmosphere many objects burn readily. In a cabin soaked in pure oxygen at greater than atmospheric pressure for five hours, almost anything burns. The most likely source of ignition lay under Grissom's couch where bundles of wires ran across the floor of the capsule. These wires had been moved, squeezed, bent, trodden on and generally abused in the process of building and re-engineering the capsule. So many engineering changes had been made over the months it took to build capsule 012 (as it was known to the contractors) that no-one had a clear idea of exactly what configuration the spacecraft was in—a fact that the investigators came to understand with some amazement.

In the pressure to get to the Moon by the end of the decade, management short cuts had been taken. Changes were being made on the hoof, and no records were being kept of what was being done and by whom. Remember also that the design of the Block I spacecraft had started before the mission mode had been finally decided. Apollo was a work in progress. Re-designs were being made as the craft were being built.

It is probable that one of the electrical wires had had its insulation damaged. A spark occurred and the inside of the capsule turned into a blowtorch.

Incredibly, with the flames surrounding them, the crew reacted calmly and exactly as they had been trained. Grissom tried activating the depressurization lever, itself in flames, to vent the cabin atmosphere. White twisted round in his seat to start opening the hatch. Chaffee lay there attempting to contact the controllers and then turned to help White. Inside the command module metal pipes melted and dripped onto the floor. Opening the hatch under such conditions was a physical impossibility.

The design of the side hatch was actually in three parts. The inner hatch opened inwards, then there was an outward-opening outer hatch and finally a third hatch that was part of the boost cover[11]. Under normal circumstances this combination could not be opened in less than ninety seconds, more likely two minutes.

The crew died within seconds of the fire starting. A full, and quite harrowing, account of the events of that day can be read in *Moon Shot* by Alan Shepard and Deke Slayton.

In the aftermath, confidence in the whole Apollo programme was severely dented. The nature and wisdom of the race to get man to the Moon was questioned and the programme faltered. Heads rolled both in the NASA structure and in the contractor's management.

The start of the recovery was probably the publication of NASA's internal report. There had been criticism directed at NASA and Congress for not holding an independent inquiry. In the end people marvelled at how frank the NASA report was. Just as one single source of ignition within the command module could not be isolated, no single person responsible for the disaster could be blamed.

The problem was one of culture and pressure. For months the astronauts had been complaining about the changes in the design. The simulators in which they trained were not able to keep up with the number of changes being installed by the engineers building the real craft. Ironically, the back-up crew for Apollo 1 were sealed inside a second Apollo command module doing their own test at the time of the fire. After a period of maddening frustration with the faults on board that capsule (at one point the hatch had fallen on one astronaut's foot), the test was called off.

Aside from not being able to keep up with the changes, the astronauts found that they were unable to establish a working relationship with the engineers. Many of the crew slated for Apollo missions were experienced astronauts from the Gemini space programme. The Gemini capsule had been built by McDonnell Douglas and a good relationship between the astronauts and the engineers had made the job of preparing capsules much easier. The contractors listened to the people who would have to fly the thing and made the changes they requested.

There was no doubt that the people at North American Aviation were just as competent as those who built Gemini, but they had never built a spacecraft before. There also seemed to be no clear line of complaint that the astronauts could use that would get results. Their criticisms were not confined to North American. They found the designers working on Apollo within NASA just as unhelpful. Great frustration built up because the experienced astronauts felt that they were not being listened to. However, in the end they recognized that the deadline had to be met and there was something of a spirit in them (after all they were mostly test pilots) that said—just get the thing in the air and we will fly it.

The combination of all these factors led to the tragedy on pad 34.

But, the lessons were learned. Changes were made to the command module. The number of flammable components was drastically reduced and much more stringent tests and construction checking were carried out (the investigators who stripped down the charred Apollo 1 capsule had found a spanner lodged in part of the wiring loom). A much simpler outward-opening hatch was designed and installed along with numerous other details in what amounted to a total design review. The pure oxygen ground atmosphere was replaced with a nitrogen/oxygen mixture that was gradually switched over to pure oxygen after launch. Confidence began to return when the astronauts started to say in public that they were delighted with the changes being made. In answer to a question posed by a congressman, Frank Borman answered:

> *'You are asking us do we have confidence in the spacecraft, NASA management, our own training, and... our leaders. I am almost embarrassed because our answers appear to be a party line.*

> *Everything I said last week has been repeated by the people I see here today. The response we have given is the same because it is the truth... We are trying to tell you that we are confident in our management, and in our engineering and in ourselves. I think the question is really: Are you confident in us?'*

In the end the spirit of Apollo won. It is arguable that without the fire the command module would never have been safe to fly and that some more serious disaster would have eventually taken place. Certainly, in the rush to get to the Moon a pause was needed to gather threads together. As Grissom himself said some weeks before the fire:

> *'We're in a risky business... and we hope if anything happens to us, it will not delay the program. The conquest of space is worth the risk of life... Our God-given curiosity will force us to go there ourselves because in the final analysis only man can fully evaluate the Moon in terms understandable to other men.'*

5.3 The service module

Dimensions
Length: 7.6 m
Basic diameter: 3.9 m
Overall mass: 24 523 kg
Structure mass: 1910 kg

As its name implies, the service module was designed to provide the domestic services that the crew would need during the journey to the Moon and back. It was connected to the command module until just before re-entry into Earth's atmosphere (generally until 15 minutes before the command module hit the outer layers of the atmosphere). Electrical power, oxygen, drinking water etc were all produced or stored in the service module. It also provided the main propulsion system for the Apollo spacecraft. The large SPS (service propulsion system) engine was needed for mid-course corrections, capturing Apollo in lunar orbit and breaking free of that orbit again. The service module was also equipped with 16 small reaction control engines (thrusters) that could be used for manoeuvring the spacecraft about during docking. These were grouped into four sets of four placed at 90° intervals round the circumference of the module. Each set (or quad) contained two that fired forwards or backwards along the module's axis and a further pair that fired round the circumference either clockwise or anticlockwise. They used the same propellant as the thrusters on the command module, but they were slightly larger as they produced 450 N of thrust. The service module thrusters were also used to establish the barbecue roll that helped to keep the Apollo cool.

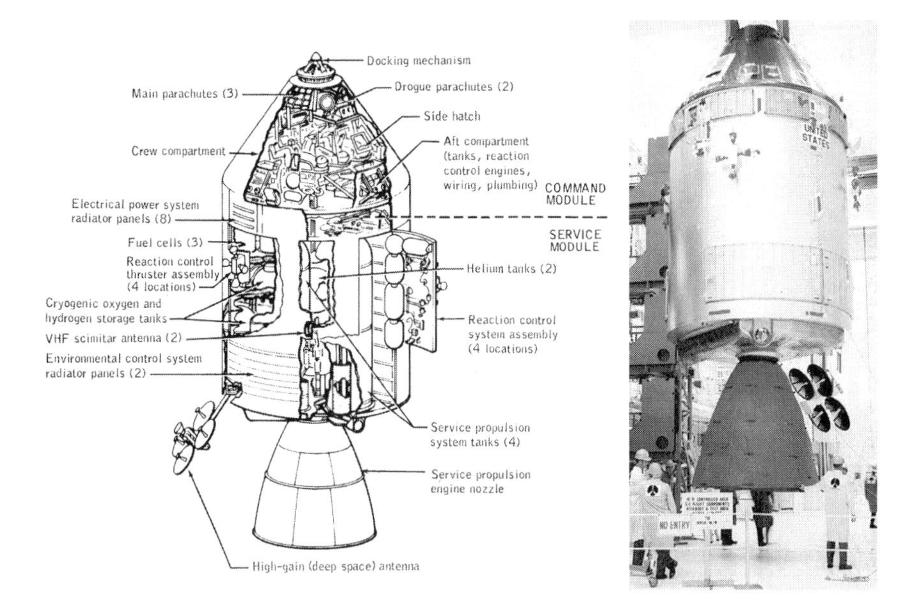

Figure 5.12 Left, the interior of the service module. Right, the Apollo 13 CSM stack being unpacked at Kennedy prior to installing it on the Saturn V. The engine bell of the SPS and the thrusters of the RCS are clearly visible. Protective covers surround the command module and the radiators on the side of the service module.

The basic structure of the service module was a cylindrical design divided into six wedges that surrounded the main engine. It was constructed from an aluminium alloy formed into a honeycomb bonded between inner and outer sheets.

A central cylinder 1.12 m in diameter contained the engine and two nitrogen tanks. Running radially from this central core were six aluminium beams which divided the module into its sectors and to which the external panels were bolted. At either end of the module were two bulkheads, and a heat shield across the bottom protected the module from the heat produced by the SPS.

On the later missions sector one contained various instruments and a camera used for scientific work. The other sectors contained various systems as follows:

- Sector 2—this portion of the service module contained part of a cooling radiator which overlapped into sector 3, a set of manoeuvring thrusters, the fuel tanks for all the thrusters, and the oxidizer sump tank for the SPS.
- Sector 3—this sector also had a thruster set on its outside panel and contained the SPS oxidizer tank. One of the two large cooling radiators

for the environmental control system (continuing from sector 2) dominated the lower part of the outside of this sector.

- Sector 4—this crucial segment of the service module contained the three fuel cells that provided electrical power for the spacecraft. Driving these cells were the contents of two liquid oxygen tanks and two liquid hydrogen tanks. The oxygen tanks also provided the atmosphere for the command module.
- Sector 5—a second environmental control radiator started in this sector and continued into the next one. This region contained the fuel sump tank. A thruster set was mounted on the outside.
- Sector 6—the SPS fuel storage tank, the remainder of the second environmental radiator and the final thruster set were all contained in this sector.

Running round the top of the service module above the bulkhead was a ring (fairing) that formed the join between the command and service modules. With its domed lower heat shield, the command module had to sit inside a shallow dish at the top of the service module. Mounted round this fairing were the eight radiators used to vent the excess heat produced by the fuel cells into space. Three tubes (one for each fuel cell) ran through each radiator.

In the vacuum of space, energy cannot be conducted or convected away. The only mechanism that works is radiation. The two larger (2.7 m^2 each) radiators round the lower part of the service module carried a water/gylcol mixture from the command module where it had absorbed heat from the cabin and the electrical systems there. Passing this liquid through pipes in the radiator allowed it to radiate energy into the cold of space. The coolant could then be circulated back into the command module.

The service propulsion engine itself was just over 1 m long, but its engine bell extended 2.7 m out of the base of the service module and opened out to a 2 m maximum diameter. The engine had to be very reliable with the smallest number of parts possible. If it did not light at the appropriate moment the spacecraft could not enter lunar orbit, or worse, break free of that orbit to return to Earth. For the sake of reliability, pumps were not used to transfer fuel and oxidizer to the thrust chamber. Instead pressurized nitrogen in the two tanks mounted in the central core was used to force fuel and oxidizer from their storage tanks into their respective sump tanks and from there into the thrust chamber.

A hypergolic propellant mixture (fuel 50% hydrazine and 50% unsymmetrical dimethylhydrazine (UDMH), oxidizer nitrogen tetroxide) was used for reliable ignition. 75% of the module's mass was the propellant.

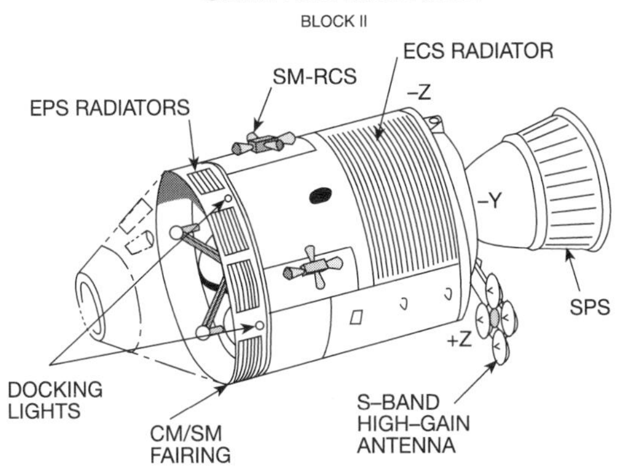

Figure 5.13 The service module showing the electrical power system (EPS) radiators, an environmental control system (ECS) radiator, some service module reaction control system (RCS) thrusters, the service propulsion system (SPS) and the main communications aerial.

The SPS provided 91 000 N of thrust, could fire for a maximum of 8.5 minutes and be restarted 36 times. The propellant mixture had a specific impulse of 314 seconds and the SPS could provide a ΔV of 2.8 km s^{-1}.

Fuel cells (electrical power system—EPS)

The electrical power to the command module was provided by the three fuel cells in the service module. Each fuel cell (manufactured by United Aircraft Corporation's Pratt & Whitney Aircraft Division) consisted of 31 separate cells connected in series. The sub-cells produced electrical energy via a chemical reaction between oxygen, hydrogen and the electrolyte liquid in the cell.

Each fuel cell provided 28 V and was rated at 1.5 kW. The three operated in parallel although only one cell was required to ensure the safe return of the spacecraft (mission rules, however, prevented a lunar landing if only one cell was working on the service module). The command module needed about 2 kW of electrical power, which is an impressively low figure—equivalent to that of many microwave ovens today.

The sub-cells comprised a hydrogen compartment with an electrode made from nickel and an oxygen compartment with its electrode made from a mixture of nickel and nickel oxide. The electrolyte solution, a mixture of potassium hydroxide and water, provided a path to allow electrical conduction between the two compartments. At the hydrogen end, hydrogen gas from the cryogenic storage tanks reacted with hydroxide ions in the solution, a process that was catalysed by the electrode[12]. In chemical terms the reaction was:

$$H_2 + 2OH^- \rightarrow 2H_2O + 2 \text{ electrons}$$

the OH^- being the hydroxyl ions from the potassium hydroxide solution. The electrons produced were drawn up the electrode to drive the current produced by the cell.

Meanwhile, at the oxygen end, that gas was also reacting with the electrolyte solution—once again catalysed by the electrode. In this reaction electrons were drawn down the electrode which then formed the return conduction path for the current from the electrical systems:

$$O_2 + 2H_2O + 4 \text{ electrons} \rightarrow 4OH^-.$$

In addition to providing electrical energy, the two reactions produced heat and water as by-products. The warm water was piped to the command module where it could be used for washing and added to dehydrated food packages. The heat was used to keep the electrolyte solution at the correct temperature, and the rest radiated into space from the radiators around the top of the service module.

The tanks used to store the oxygen and hydrogen for the fuel cells had to be kept at cryogenic temperatures in order to hold the gases in their liquid state[13]. The tanks had most impressive design parameters. For example:

- very low heat loss to the surroundings—an Apollo cryogenic tank could keep ice frozen for over eight years if placed in a room heated to 20 °C;
- very low leakage of gases out of the tanks—the same leakage rate in a typical car tyre would flatten the tyre in 32 400 000 years.

The most daring mission—Apollo 8

The period between 1967 and 1968 was not the most encouraging as far as NASA was concerned. The aftermath of the Apollo 1 fire had left many questioning the basic concept of an attempted Moon landing. The Saturn V rocket flew, but not without some teething problems, and the lunar module's development was not proceeding as quickly as everyone had hoped. In the background the Vietnam War rumbled on, consuming more and more American resources and morale—which had already been badly sapped by the assassinations of Martin Luther King and Robert Kennedy.

As 1968 drew to a close the Apollo 7 command and service module stack sat atop a Saturn Ib rocket ready for a manned Earth-orbital test of the systems in

October. Apollo 8 was scheduled to be the first Earth-orbital test of the lunar module. However, delays in delivering a working LM to NASA probably meant postponing that flight until the New Year—putting the end of the decade target in some doubt. As August came round the idea for a daring re-design of Apollo 8 started to dawn. If Apollo 7 flew well, and if the lunar module could not be ready for the planned Apollo 8, then why not fly a new mission—to the Moon. The Saturn V booster would need to fly a manned mission at some point, and as it had the capability to fling the command and service modules to the Moon why use it for another Earth-orbital flight? The idea was the brainchild of George Low, the NASA engineer with responsibility for overseeing the development of the Apollo spacecraft. Predictably, the early reaction among some of the NASA management was not enthusiastic. For one thing the guidance systems (including the software) for a Moon flight had not yet been fully worked out. When the NASA administrator James Webb first heard the idea he was at the American Embassy in Vienna. He yelled over the transatlantic phone line 'are you out of your mind!' The risk was enormous. To send only the second manned Apollo flight to the Moon without a lunar module (always considered as a back-up lifeboat in case something went wrong) could potentially jeopardize the whole NASA organization.

However, there was another factor to consider. According to the CIA the Soviets were about to resume launches of their Soyuz spacecraft (after the disastrous flight of Soyuz 1 culminating in the death of Vladimir Komarov when his craft crashed into the ground after re-entry). Although intelligence was reasonably sure that they could not land a man on the Moon before the end of the decade (they had not tested any boosters powerful enough), they were capable of sending a spaceship on a loop round the Moon. If they did that, then the Soviets could reasonably claim to have beaten NASA to the Moon.

In fact, the Soviets were planning an ambitious programme called Zond. This involved flying several tanks of fuel into Earth orbit, where they would be combined with a small manned craft for a landing on the Moon. The plan faltered when Komarov died. Rather than abandoning it completely the plan was re-structured as an attempt to orbit the Moon before the Americans. In September 1968 they flew an unmanned Zond round the Moon, and in November of the same year another craft, loaded with tortoises, flies and worms, took the same trip. Then, for once, the Russians hesitated. They wanted more tests. The cosmonauts wanted to go, but they were held back. The Americans might still fail—it was better not to risk all so soon.

Within the planning teams at NASA, the response to the prospect of sending a crew to orbit the Moon was more enthusiastic. Webb could see the political pay-off both abroad and at home (Congress needed to see a major NASA success to ensure the continuity of funding) and had agreed to give the idea

Figure 5.14 Earthrise over the Moon as seen from Apollo 8.

careful consideration. When the engineering staff (including Von Braun and the Marshall team) said that they could fly to the Moon by December, he sanctioned the idea—provided Apollo 7 flew without any problems.

On 11 October 1968 Wally Schirra, Donn Eisele and Walt Cunningham lifted off for an eleven day Earth-orbital test of the Apollo command and service module combination. This mission has in many ways been eclipsed by the more spectacular achievements of the ones that followed, but that is a shame. The mission was a total success and the Apollo spacecraft performed almost flawlessly. Without that major shot in the arm, the Apollo programme, and Apollo 8 especially, would have been set back considerably.

Apollo 8 lifted off for the Moon on 21 December 1968. Sixty six hours later the spacecraft slipped behind the Moon on its free return trajectory, the crew inside (Frank Borman, Jim Lovell and Bill Anders) readying themselves for the SPS burn that would slow them into lunar orbit. At the same time, they became the first men to see the reverse side of the Moon with the naked eye. This was a mission of firsts. On the way the Saturn V third stage had boosted them to a speed faster than any human being had travelled before. They were also the first men to view the Earth from such a distance that the entire globe could be seen at once.

Christmas Eve 1968 saw the most extraordinary TV broadcast in the history of mankind. The crew of Apollo 8 caught Earthrise above the Moon on camera for all to see.

Historically it was a moment of supreme significance. The American people were involved in a bloody war—and many of them were questioning its morality. Civil rights leaders were struggling to come to terms with the death of their icon and many were still astounded by the assault on democracy implied by the assassination of a President. Yet here was a view of the Earth that showed no political, racial or geographical boundaries. Furthermore, the crew had chosen to end their broadcast by reciting in turn lines from the Book of Genesis:

> 'In the beginning God created the heaven and the Earth; and the Earth was without form and void . . . '

It is impossible now to convey the impact that this had on the listener.

The next milestone was the SPS burn to inject the crew back into a free return path. Once again, that had to be carried out behind the Moon. The tension in the crew was evident to those who listened:

Lovell: 'did you guys ever think that one Christmas Eve you'd be orbiting the Moon?'

Anders: 'just hope we're not doing it on New Year's.'

After the burn had been successfully accomplished Lovell greeted ground control once the spacecraft emerged from round the Moon with the following exchange between him and the CAPCOM[14] (Ken Mattingley):

Lovell: 'Houston, Apollo 8. Over.'

Mattingley: 'Hello Apollo 8. Loud and Clear.'

Lovell: 'Please be informed there is a Santa Claus.'

Mattingley: 'That's affirmative. You are the best ones to know.'

The lasting achievement of NASA's most daring mission was one that nobody expected—a new perspective on the Earth. Many people have dated the start of the environmental movement to the first sight of the Earth rising over the lunar highlands. In the words of Bill Anders,

> 'we flew all that way to explore the Moon, and the most important thing we discovered was the Earth.'

The almost tragedy—Apollo 13

The Apollo 13 mission in many ways represented NASA's finest hour during the Moon exploration programme. This sounds ironic considering that the mission was designed to land on the Moon and instead swung round and returned to Earth without orbiting. However, consider the brilliant improvisation that went on, the endurance (both mental and physical) shown by crew and mission controllers, the way in which various systems were pushed to their design limits and functioned perfectly, and the tremendous support expressed by people around the world. After an explosion had deprived the crew of electrical power and oxygen there was a very real chance that they would die. Yet they were brought home safely by the combined efforts of the whole NASA team.

The story of the Apollo 13 accident starts before the Gemini programme had been completed. The command and service module contractors, North American Rockwell, had subcontracted the construction of the oxygen and hydrogen tanks, used to store the liquids needed to feed the fuel cells, to Beech Aircraft. The contents of the tanks were normally kept at −207 °C, at which temperature they were a rather slushy consistency. However, just enough of the oxygen and hydrogen would evaporate off to feed the pipes running to the fuel cells and, in the case of the oxygen, the command module cabin. Occasionally it was possible that the pressure in the tanks would drop, and so heating and stirring systems were incorporated.

Early in the service module design the operating voltage was fixed at the 28 V supplied by the three fuel cells. This information was passed on to Beech and they designed the heaters and thermostats for the tanks to operate at 28 V. However, it was realized that while sitting on the launch pad the service module would be running on external power at 65 V. When North American became concerned that this would cause problems in the tanks they instructed Beech to up-rate the components. For some reason, the thermostat switches were never changed to 65 V versions. NASA, North American and Beech all reviewed the designs and the mistake was not caught. All the Apollo service modules up to 13 flew with the wrong thermostat switches.

The next step in the chain leading to an explosion came once the tanks had been delivered to North American for installation. The tank that eventually found itself as oxygen tank 2 on Apollo 13 started its life in Apollo 10. However, due to a re-design the tank was removed and replaced with a newer version. It was modified and then installed on the service module for Apollo 13. During the removal process one of the bolts holding the shelf on which it was mounted in sector 4 of the service module was not removed. The crane tried to lift the shelf. The shelf moved upwards and then fell back with a small thump. The tank was inspected for faults and found to be fine, so it was modified and fitted to the service module for Apollo 13.

Unfortunately there was a problem with the oxygen tank that was not discovered until the countdown test. During this test all the propellants were loaded on the spacecraft and the crew sat in the command module for a full rehearsal. At the end of this the propellants were extracted from the ship. The procedure for

Figure 5.15 The Apollo 13 service module after being jettisoned four hours before re-entry. The damaged section can clearly be seen in the lower right of the picture.

emptying the tanks was to pump pressurized oxygen in that would force the liquid out of the drain pipes. This was not happening to tank 2. For eleven days the engineers studied the problem, eventually surmising that one of the drain lines had been jolted so that oxygen was leaking from the inlet directly to the drain and not forcing the liquid anywhere.

Annoying though this was, it was not a threat to the mission. In flight the drain line would never be used. As long as they could get the LOX out now, there

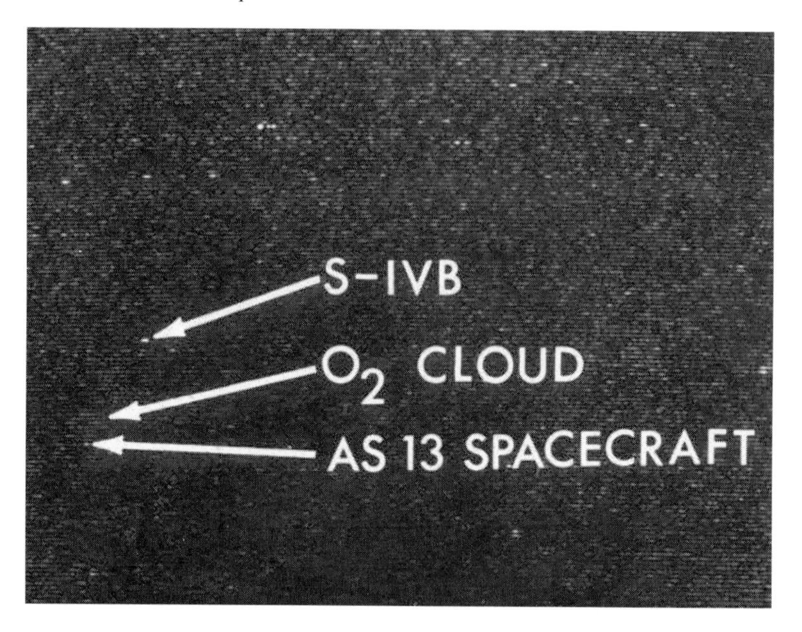

Figure 5.16 This remarkable photograph taken from Earth shows the Apollo 13 spacecraft surrounded by a sphere of debris. The Saturn V third stage is also visible.

would be no problem on launch day. The solution they came up with was to turn on the heaters. Boiling the liquid oxygen was as good a way of getting it out of the tank. Considering the time it would take to remove the faulty tank, replace it and check out the new one, it was decided not to risk a delay to the launch, and to heat up the tanks.

However, what they did not realize was that the thermostats, designed to turn off the heaters at 27 °C, were only rated at 28 V. Estimating that it would take eight hours to boil off the LOX, the engineers threw the switches to run the heater in tank 2 from the external power supply. When the time came for the thermostat switches to open, the higher current at 65 V fused them shut. The temperature in tank 2 rose to something like 500 °C. The ground technicians were unaware of this. As the design called for the thermostats to cut the power to the heaters at 27 °C, their gauges did not go any higher than that temperature! Inside tank 2 the Teflon insulation round the cables leading to the fan motor melted.

Seventeen days later and 200 000 miles (322 000 km) out into space, the command module pilot Jack Swigert threw the switch to stir the contents of oxygen tank 2.

GRUMMAN AEROSPACE
BETHPAGE, NEW YORK

SOLD TO: North American Rockwell
Space Division
Downey, California

Invoice No.
70-417

Invoice Date	Waybill no.	Items Net
17 April 1970		For Bethpage, N.Y.

Item	Quantity	Description	Unit Price	Amount
1.	400,000 mi.	Tow, $4.00 1st. mile $1.00 ea added mile Trouble call, fast Battery charge (road call + $.05 KWH)		$400,004.00
2.	1 KWH	Customer's jumpers		4.05
3.	50 lbs.	Oxygen at $10.00/lb.		500.00
4.	1	Sleeping Accom. for 2, no TV, AC, with radio, modified American plan		Prepaid
5.		Extra quest $8.00/ night		32.00
				$400,540.05

Figure 5.17 The 'towing bill' sent by Grumman (constructors of the lunar module) to North American (command and service module) after the safe return of Apollo 13.

Up to this time, the tank had been full enough to cover the wires in liquid. Now they were exposed. A spark flew, igniting the gaseous oxygen in the tank. The pressure build up blew the top of the tank off. Exposed to vacuum, the contents of tank 2 evaporated in a flash, filling sector 4 of the module. The rapid pressure build up blew the side of the sector clean away from the spacecraft. The other oxygen tank was undamaged, but as it shared some pipe work with tank 2 it also found a path to space and started to leak. The explosion also jammed some of the valves running to the thrusters. When the ship was rocked by the explosion, the autopilot started firing the thrusters in an attempt to steady the ship. Some of them did not fire, so the autopilot became horribly confused, continually hunting for a combination that would steady the ship.

With the fuel cells fed by tank 2 dead and the other dying, and with the oxygen needed to breathe in the command module visibly venting into space, the crew faced an imminently fatal situation. The solution was to transfer to the lunar module and live in that for the remainder of the flight. In a frantic few minutes the crew had to power up the LM, transfer guidance control from the CM to the LM and shut down all the electrical systems in the CM in an effort to preserve the batteries needed to power the spacecraft during re-entry.

The rest of the mission became a story of survival.

- The LM descent engine was used to provide the ΔV needed for the transfer to a free return trajectory. With all electrical power turned off, the cabin temperatures dropped to those of a meat storage freezer. Drinking water froze. The crew were instructed to stop dumping their urine overboard as the crystalline cloud that it formed round the spacecraft was disturbing the tracking information from Earth.

The complete story of Apollo 13 has, of course, been immortalized in film, but it can also be read in Jim Lovell's own book (written with Jeffrey Kluger).

After the investigation into the accident had issued its report, recommendations were made to alter the design of the spacecraft to prevent such an occurrence in the future. Apollo 14 was modified to include an additional oxygen tank, a back-up battery in the service module, and the removal of the fans and thermostat switches from the oxygen tanks. The command module also had provision for storing an emergency water supply added.

Notes

[1] Once the decision to use docking on the mission had been taken, the Gemini programme was used to gain experience of the manoeuvres required.

[2] In December 1966 NASA decided that there would be only one Block I spacecraft flight—Apollo 1. When the crew of that mission died in a fire during a test the widows asked that Apollo 1 be reserved to designate the mission that never flew. As mission designs changed, so did the numbering system, and so the first manned Apollo command module flight was Apollo 7. The first flight of the Saturn V, with a basic Block I command module that incorporated many of the Block II features, flew on 9 November 1967 as Apollo 4.

[3] A honeycomb provides the most rigid structure with lightest weight. You can make a simple honeycomb by taking several straws and gluing them together side by side. If the straws are then cut to 2 cm length and card glued to the top and bottom (at right angles to the length of the straws) then a very lightweight but rigid structure is produced.

[4] The space shuttle uses specially designed ceramic tiles with very high thermal resistances to prevent conduction during re-entry.

[5] The Pascal (Pa) is the SI unit of pressure. 1 Pa is the equivalent of a 1 N force acting over 1 m^2. Normal atmospheric pressure is about 100 kPa, so the cabin pressure was just over one third of this. The decision to use this low pressure was partly based on engineering constraints. With a much higher pressure it would have been difficult to

re-stock the cabin atmosphere in flight. The mission design called for the possibility of the astronauts having to space-walk to cross from the LM to the command module should the two be unable to dock. If this was to be done without needing an air-lock (which would add weight and complexity) then a low cabin pressure was required.

[6] This became a problem during the Apollo 13 flight. With the service module crippled by an explosion, the astronauts were forced to live most of the time in the lunar module. This put great demands on the LM's atmospheric recycling unit, which had been designed for only two occupants. The CO_2 filter rapidly became saturated and there was a danger of the crew suffocating on their own breath. Unfortunately, the filters for the command module were a different shape (square) to those used in the lunar module (round). The mission controllers at Houston had to rapidly come up with a method for fitting a square object into a round hole in order to save the lives of the men. The improvised device had to be constructed from cardboard flight manual covers, tape and various other items scrounged from spare parts on the ship.

[7] During the Apollo 13 disaster, the astronauts had to turn off all the heating equipment in the command module as a means of conserving electrical power. The temperature dropped to that of a meat storage freezer. While attempting to sleep in the command module they discovered that their bodies could warm a thin layer of air surrounding them provided they did not move and disperse it. This would not be possible if convection was taking place.

[8] On the ill-fated Apollo 13 mission the command module pilot, Jack Swigert, was instructed to throw a switch that turned on the fan in one of the liquid oxygen tanks. This was a routine request from mission control as the fan was responsible for stirring the tank to prevent settling of the liquid. Unfortunately, in this case there was a fault associated with the wiring to the motor. The result was an explosion that crippled the service module and very nearly cost the lives of the crew.

[9] This happens because the air travelling over the top surface of the wing is moving faster than that over the bottom surface. The faster the flow rate of a fluid, the less the pressure it exerts (see chapter 3). Consequently, the pressure below the wing is greater than that above it, producing a net upward force. As the command module entered the atmosphere tipped backwards so its heat shield was protecting the crew, the same effect happened.

[10] The main report was a 20 cm thick, 3000 page stack of 14 booklets.

[11] This complex system had been designed to minimize the risk of the hatch failing in the vacuum of space. With multiple hatches redundancy was assured, and with an inner hatch that opened inwards the designers could be sure that the pressure inside the capsule would force the hatch shut. The astronauts were happier with a much simpler arrangement that ensured that they could get out in a hurry.

[12] A catalyst is a substance that allows a reaction to take place, but which is not itself consumed in the reaction. Many chemical reactions will either not take place at all, or only very slowly, without the aid of a catalyst.

[13] A given mass of gas can be stored much more effectively as a liquid as it takes up less volume.

[14] CAPCOM—the capsule communicator—an astronaut responsible for the direct radio communications between Earth and the spacecraft. No instructions were allowed to be relayed except via the CAPCOM.

Intermission 3

Inertial guidance and computers

The information in this section comes mostly from James E Tomayko's paper on computers used by NASA—this is available on the web at the address listed in the bibliography at the end of the book.

I3.1 The need for a guidance system

Imagine going on a journey for which there are no signposts or maps to tell you where you are at any moment. The only landmarks that you can use are either so large that they give you no accurate position, or so small that you need a telescope with cross hairs to line up on them.

How can you tell where you are?

This is the problem faced by astronauts travelling to the Moon. There are very large landmarks—the Earth, the Sun and the Moon—that can be seen, but as they are so large it is difficult to use a telescope to provide a precise angular fix on where you are. There are other landmarks, which are small enough for an accurate fix (e.g. the stars), but it is a very complicated task to navigate to the Moon entirely on the basis of star sightings.

The answer to this navigation problem is to use the principle of *inertial guidance* first developed by the Germans as a means of guiding Nazi missiles during the war. The principle of inertial guidance is very simple. You can always tell where you are relative to an established starting point by keeping a precise track of the speed and direction in which you are travelling. If you start from your house and walk at 3 m s^{-1} due south for 5 minutes, then turn 90° left and walk at 5 m s^{-1} for 10 minutes, you should be able to tell precisely where you are with a little bit of calculation. A spacecraft has to do this in a slightly different way. It needs to know how long its engines have been burning for, the ΔV that they achieved and the angle along which the thrust was directed.

From this the computer can work out the velocity change, the direction in which the spacecraft is moving and how far it has travelled along that direction. Consequently it can keep track of where it is.

At first the mission planners thought that all the guidance and control could be performed on board by the combination of the ship's computer and astronauts acting as navigators. However, the task proved to be so complex and took up so much space in the computer's memory that a switch to ground-based guidance was made. Even so, the system on board the spacecraft retained some independence should contact with ground control be lost.

Trajectory monitoring from the ground was done by a sophisticated system that measured the shift in the radio frequency being used to communicate with the ship. This *Doppler* effect is frequently heard on the ground when a fast-moving ambulance or police car passes the listener. As the vehicle approaches, the pitch of its siren increases and then it decreases as the vehicle speeds past. The same effect occurs for radio waves produced by a moving object. The systems developed for Apollo were so sensitive that they could even detect the slow rotation (barbecue roll) used to help keep the temperatures down on board. Computers on the ground were able to predict the precise Doppler shift that should be occurring at any moment on the pre-defined trajectory and compare it to what was being measured in order to check the spacecraft's path.

I3.2 Guidance and control systems

On Apollo the guidance and control system was designed to perform several functions:

- Calculation of the spacecraft's position and velocity (together this information was known as the *state vector* of the spacecraft).
- Optical checks of the spacecraft's navigation platform.
- Control and measurement of the spacecraft's attitude (the direction in which it was pointing).
- Control of the propulsion system.
- Control of the path followed by the command module during re-entry.

Central to these functions was the *inertial measurement unit*. This was a solid block of machined beryllium, gimbal-mounted[1] and gyroscope-stabilized, which formed a stable platform for an array of sensors that measured accelerations and rates of rotation about three axes at right angles to each other. The platform was locked down until shortly before lift off, when it was released to move relative to the spacecraft. Given Newton's first law of motion, the stable platform in fact tended to maintain a fixed orientation as the spacecraft rotated about it.

This information, along with data about accelerations and rotations, was passed to the computer for use in tracking the spacecraft's attitude and trajectory.

Periodically it was necessary to check that the stable platform had not drifted away from its initial orientation. Even small errors in the information passed to the computer would produce large mistakes in the spacecraft's trajectory over the distances involved in flying to the Moon. With the trajectory designed to fly them round the far side of the Moon within 60 miles of the surface, the crew were understandably keen to have an accurate fix. This was done by using the computer to rotate the spacecraft and angle the ship's telescope (or sextant) to point to a given star or alternatively a landmark on the Moon. With this done the navigator (command module pilot) would have a look through the optical equipment. Invariably the target star would not quite be centred in the sights, in which case he would record the angular error which was then passed on to the computer.

On the first orbital mission, Apollo 8, Jim Lovell was responsible for taking the sightings that helped align the inertial system. While in orbit about the Moon he took fixes of various lunar landmarks to help define the ship's orbit and then to take reference marks for the landing to come. In this mode the computer acted as an automatic aiming system. All he had to do was feed in the coordinates of the next crater or mountain and the sextant would be automatically positioned. The computer even kept track of the ship's motion so that he did not have to adjust the aim as they flew past the landmark. On his later Apollo 13 mission Lovell was faced with a serious problem when the explosion on board the service module scattered debris in a huge sphere about the spacecraft. With the various shards of metal glinting and reflecting in the sunlight is was impossible to tell if he was looking at stars or shattered remains of the ship. Eventually the problem was solved by using larger references—such as the Sun.

Although the platform was given a very sophisticated mounting it was not able to have complete rotational freedom. If two of the gimbals supporting the platform happened to line up, then the axes locked and the guidance platform would be rendered useless. The cabin instruments included an 'eight ball' or artificial horizon on which were painted red circles showing when the system was close to gimbal lock. Under normal circumstances the spacecraft would not be manoeuvring so violently that this would be a problem. However, in the aftermath of the Apollo 13 explosion the computer was trying to compensate for the venting oxygen which was thrusting it off alignment. It was doing this by firing the thrusters on the service module. Unfortunately not all of them were working, causing some violent movements. The same problem arose later when the crew had transferred over to the lunar module. Lovell tried to use the LM's thrusters to manoeuvre out of the cloud of debris that was preventing him from getting a star fix. However, the lunar module's computer and thruster systems were calibrated for moving the LM about on its own—not with a very large and dead command and service module combination sitting on its back. The resulting motion nearly sent the LM's inertial platform into gimbal lock.

An inertial platform check was generally performed prior to an engine burn. One of the functions of the computer was to control the burn activation, duration and direction of thrust. In some cases this was quite critical—some burns had to be very accurately timed. A couple of seconds too long could mean crashing into the Moon.

Directing an engine burn is not simply a matter of pointing the whole spacecraft in a particular direction and firing the engines. The service module's main engine was gimbal mounted, so the direction of thrust could be different to the direction in which the craft was pointing. This degree of control was necessary as the spacecraft's centre of gravity moved while the propellant in the service module was used up. The engine had to gimbal to compensate for this effect.

The spacecraft's thrusters could also be controlled by the computer. These were used to rotate the ship about any one of three axes, thrust forwards or backwards (either during docking or sometimes to 'trim' a major burn that had been performed by the main engines), or maintain a set attitude or rate of rotation (such as the barbecue roll). The lunar module, service module and command module were all provided with their own set of thrusters, although those of the command module were not used until re-entry.

13.3 The Apollo computer

The need for Apollo to carry an on-board computer was recognized early in the mission planning, although the reasons listed are quite illuminating as to the state of mind at the time:

- to avoid hostile jamming(!);
- to gain experience needed for the interplanetary (e.g. Mars) missions that were to follow;
- to prevent the ground-based communications systems from becoming overloaded should there be many spacecraft flying at once.

In reality the complexity of building and flying a spacecraft to the Moon resulted in a severe downgrading of the on-board computer's role. All the major trajectory and burn calculations were performed on the ground and the solutions transmitted up to the crew for entry into their computer.

The contractor chosen for the Apollo computer system was the MIT Instrumentation Lab[2]. They had had substantial experience designing systems for the Polaris missile and other aerospace computers. Despite this expertise the Apollo computer turned out to be a genuine challenge that stretched the limits of the available technology.

Once the mission mode had been decided it was clear that two independent systems would be needed—one for the command module and one for the lunar module. These two primary computers were identical in design, but were equipped with different software. Normal manoeuvres carried out in flight could be backed up by the ground-based computer systems, but this did not apply to the landing. The time delay between sending a radio signal and receiving it at the spacecraft was long enough to make the difference between crashing into the Moon and a soft landing. For this reason the lunar module was equipped with a back-up or abort computer. Its only job was to monitor the functioning of the main LM computer and to control the abort back to lunar orbit if things should go wrong.

Early designs of the Apollo computers used transistor-based circuitry; however the disadvantages of this design led the team to be more daring when they came to the later designs. As early as 1962 the MIT group began to look at the newly developed integrated circuits (ICs) for use in the Apollo computers. At the time ICs were only three years old and so had no track record of reliability (always a major factor in NASA designs). MIT had to convince NASA that the advantages of using such devices were great enough to be worth the risk.

The sort of integrated circuit being used was a pale shadow of the microprocessors that are manufactured today. These ICs were single logic gates. MIT chose to make the circuits from three input NOR gates[3]. It would have been simpler to choose a selection of different logical gates, but by fixing on a single gate overall simplicity and reliability improved. Nearly 5000 gates were used in the Apollo computer. To put this in context, something like 60% of the total US production of ICs were being used on Apollo prototypes by the summer of 1963.

The final computer was housed in the command module lower equipment bay (see chapter 5) and measured 61 cm by 32 cm by 15 cm with a mass of 32 kg. The equivalent computer in the LM was placed in the cabin mid-section. All the circuits were placed in two trays of 24 modules. Each module consisted of two groups of 60 flat circuit boards with 72 pin connectors at one end. Each circuit board held two ICs. Tray A held the logic circuits, the power supply and the interface electronics that connected the computer with other systems. The second tray, tray B, had the memory circuits, the computer clock[4] running at 1 MHz (contrast that with the latest Power PC and Intel chips running at 450 MHz or more), and the various circuits responsible for setting visible and audible fault alarms for the crew. Each unit was hermetically sealed.

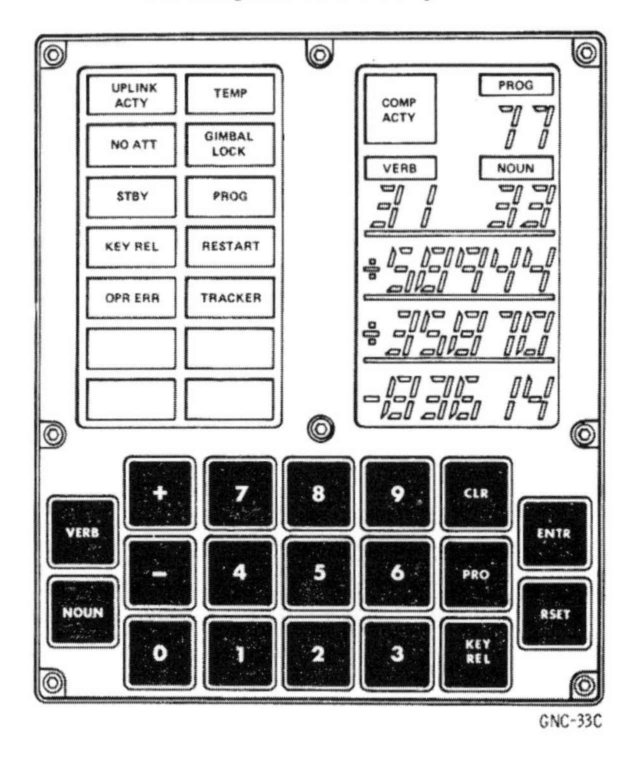

Figure I3.1 The standard display and keyboard unit used for the Apollo computers. The black squares at the bottom are keys that can be pressed. The set of rectangles at the top left are alarm and other activity lights. Those on the right are the computer displays that show the current program running, contents of memory locations (registers) and other information.

The crew could type commands and read data from a set of identical display and keyboard units (DSKYs, pronounced 'diskies').

There were two of these units in the command module—one on the instrument panel above the command module pilot's couch and the other at the navigation station in the lower equipment bay. One DSKY was mounted on the lunar module control panel. The system for communicating with the computer seems very crude by today's standards. For example, there were no alphanumeric keys available. On a modern computer a request for a program or operation would be communicated to the computer by mouse or typing a command (or shouting at the computer). On Apollo a specific request was defined by the VERB, NOUN combination. Pressing the VERB button on the keyboard set the computer to read the next two digits typed as a base 8 code for a certain operation. NOUN set the computer to read the next two digits as the base 8 code for the component on which the operation set by the VERB was to be carried out. ENTR performed rather like the modern 'enter' or 'return' key. The combination of a specific VERB and NOUN set a given computer program running. If the program required data to be entered by the crew, then the VERB and NOUN lights started to flash until every number required had been keyed in. The crew were provided with a manual that listed all the combinations of VERBs and NOUNs as well as error code translation and checklists.

As an example of how the computer might be used, VERB 3 7 ENTR instructed the computer to stop what it was doing and change to a new program. The next key presses (e.g. 3 1 ENTR) would switch to the program (P31—the rendezvous targeting program). Various requests could be made during the program:

VERB 15 NOUN 18—display manoeuvre angles
VERB 06 NOUN 84—display velocity change for next manoeuvre.

Memory requirements for the computer leapt as the amount of data and programs grew in complexity. The final configuration of the machine had 576K of RAM and 32K ROM[5]. Neither of these figures would come anywhere near running a modern operating system (which in my view says more about the modern code than it does about the computer that could fly to the Moon!). The type of memory was also very different from today's computer chips. Computers of the time used *core memory* that relied on electromagnetic principles.

The principle of core memory

A memory core consists of many small 'donut' shapes of a material called *ferrite* (a ceramic mix with iron). Ferrite is easily magnetized, provided a coil of wire wrapped round the donut has an electric current passing through it that is higher than a fixed minimum value.

Figure I3.2 shows a single memory core in what would have been a large array. There are two wires, labelled X and Y, which are responsible for magnetizing the core. Half the required current is sent along each. These wires are linked to other cores in the array, forming a set in rows and columns. The correct core can then be 'addressed' as it will be the only one which has the required current to magnetize it (half from each wire).

The core can be magnetized in either a clockwise or an anticlockwise direction depending on the sense of the current in the X and Y wires.

In this way each individual core can be magnetized clockwise or anticlockwise, recording a one or zero. Reading the memory is a case of sending a new

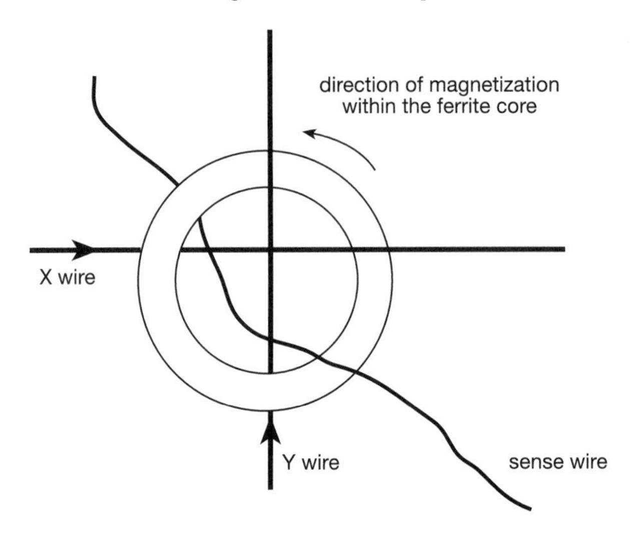

Figure I3.2 A single memory core. The current in the X and Y wires produces a magnetization as shown in the core. Reversing both currents would reverse the sense of the magnetization.

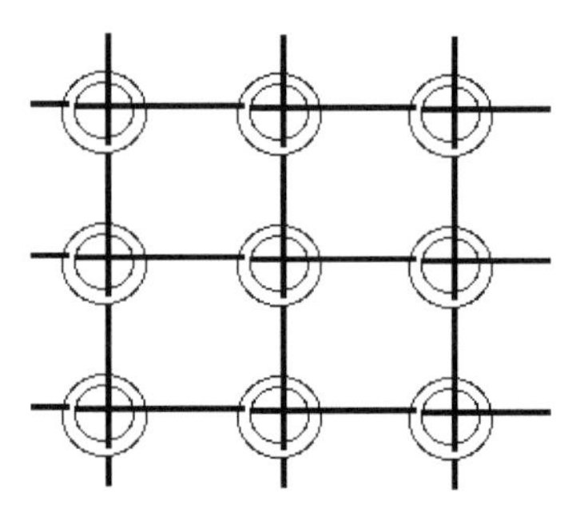

Figure I3.3 Linking cores into an array. The sense wire would wind from one core to another.

current to an individual core via the X and Y wires. If the current is in the same sense as that which magnetized the core, then nothing happens. However, when the current in the read cycle reverses, this will re-magnetize the core in the opposite direction. The principle of electromagnetic induction states that a changing magnetic field will induce a current in a wire. As the core magnetization flips it induces a pulse of current in the sense wire. If the flip is from clockwise to anticlockwise, the current pulse is in one direction. If the flip is from anticlockwise to clockwise then the current pulse is induced in the opposite direction. Sensitive electronic amplifiers monitor the sense wires, read the current pulse direction and convert this into a 1 or 0.

The key thing here is that the *change* in magnetization is read by the sense wire—the process of reading the memory automatically erases it.

The Apollo computer's RAM was constructed of memory core of the type described above. However the ROM used a slightly different principle. Each ferrite core recorded not a single 1 or 0 (bit) but a whole 16-bit computer word. In this case each core had multiple sense wires passing through it. Each 16-bit word was represented by 16 wires. If the word was 1001000100010010 then the first wire would miss the core, the second go through it, the third and fourth miss, the fifth pass through etc (read from right to left). Now, when the memory was read there would be a current pulse only in those wires that passed through the core. Any pulse was regarded as a 1 and the absence of a pulse a 0. In this way each core could represent a word (in fact on Apollo each core could have four separate words recorded). The advantage of this sort of memory store (called a *core rope*) was that the memory relied on how the wires were strung and so it could never be erased—even by a total power failure. It also allowed a great deal of information to be packaged in a very small size. The difficulties were that it was very complex to build and impossible to correct should an error have been made.

Software engineering on the hoof

As a further example of how 'cutting edge ' the Apollo computer was for the time, Raytheon—the contractor chosen to build the computer equipment (MIT was contracted for its design)—increased its production staff from 800 to 2000 to cope with the demands of building 57 computers and 102 DSKYs.

However, if the computer design was remarkable for the time, then the software and its engineers were no less so. On the Apollo 14 flight a fault developed in the lunar module. During the time between separation from the command and service modules and starting the powered descent to the Moon, the crew carefully checked through the systems and ran a dummy landing through the computer. The first time they tried this they discovered that at the moment when the computer should have fired the descent engine it instead triggered the abort sequence that would, if it had been a real landing, have fired them off into orbit again. Mystified, they ran the simulation again. This time all seemed to be well for several minutes until the abort kicked in again. Worried that the landing would have to be called off, the crew were forced to wait while mission control tried to piece together a solution.

It was quickly decided that the most likely cause of the problem was a small ball of solder coming loose in the crew's abort switch. Indeed they discovered that tapping the panel could clear the fault. The only way to work round this was to attempt to re-program the computer so that it ignored the signal from the faulty abort switch. This was risky, as it would have denied the crew an easy way to trigger the abort in case of trouble. Mission control had enough faith in the ability of the crew to proceed. However they had less than three hours to rewrite the computer code.

A telephone call woke Don Eyles, the MIT software expert, who threw a coat over his pyjamas and ran down the stairs into the Air Force car that had arrived. Having been rushed to his desk, he listened to the problem and started working on his computer. He rapidly completed the new code (even finding a way to retain the abort switch) which was immediately fed into the computer in mission control's LM simulator. It worked fine.

By the time the new code arrived at mission control, Apollo 14's LM was passing out of radio contact behind the Moon. Signal was re-acquired only thirty minutes before the landing would have had to be scrubbed. The crew typed in the new data with just 15 minutes to go.

13.4 The Apollo computer in perspective

In the introduction to this book I referred to a conversation between myself and a young student regarding the relative technologies incorporated in the Apollo computer and the sort of processors that we find in washing machines today. On the face of it the comparison is almost ludicrous, so great has been the progress in computer technology. However, closer inspection reveals that although modern computers are far more powerful and complex they do not necessarily represent a crushing advantage over the Apollo system for the sort of job that it was designed to do.

It is interesting to compare a modern computer with the Apollo system in a couple of key areas.

Processing power
A modern computer's Central Processing Unit (CPU) performs calculations at a far greater rate than the Apollo computer. Trajectory calculations could therefore have been done with greater speed and accuracy. However, the calculations that were done both on board and on the ground were quite adequate for guiding the Apollo spacecraft when backed up by the expert observations made by the crew. On the Apollo 8 mission, which was the first one to travel to the Moon, Jim Lovell's star fixes were completed so quickly and accurately that some of the planned course corrections were not needed.

The DS1 probe mentioned in chapter 3 is equipped with an autonomous guidance control using star and asteroid maps stored in its memory to correlate its position. A similar system on Apollo would have made the astronaut's time easier, but unlike DS1 their mission was simply to fly to one objective.

One of the other functions of the computer was to time the duration of course-correction burns and to adjust the engine's thrust direction as the burn was taking place. The extra processing power and speed would be of little advantage over the systems of the time. The space shuttle uses computers to improve the performance of the engines (as do most modern cars) and to govern the 'fly-by-wire' systems that it incorporates—all of which represent significant advances. However the shuttle is purpose-built to lift payloads into Earth orbit in a reusable manner. The same systems would have been of little use to the command and lunar modules, but the extra processing power would have been a considerable advantage during the lunar landing. Letting the computer do much of the flying during the final stages would have left more time for the crew to pick a landing spot. The LM's computer did control much of the powered descent but, as was seen on the Apollo 11 flight, was sometimes close to being overloaded.

Crew interface

Modern computers use full keyboards and very sophisticated operating systems that govern how the user interacts with the computer. Voice recognition software is becoming increasingly accurate and flexible and is now even appearing in cars to control air conditioning and telephone functions. In contrast the Apollo astronauts were faced with the complex VERB NOUN system and the error codes that the computer produced had to be looked up in a book (those that were not familiar from training, that is). In this area a modern system would have huge advantages over the Apollo computer, especially during a crisis.

The space shuttle is a much more computer-intensive machine than the Apollo spacecraft was. However its computers are rooted in the same sort of technology as Apollo (they also use core memory for example) and although the shuttle is the first spacecraft to use cathode ray tube (TV screen) type displays for information the screens are crowded and the keyboards used to enter data are similar to the Apollo ones. It has been estimated that the 13 000 keystrokes needed to fly a week-long Moon mission are equalled by one 58 hour shuttle flight. On some early shuttle flights the crews took to carrying HP-41C programmable calculators[6] to supplement the main computers—and they found them easier to use. Laptops are taken on board now.

Clearly there is still some way to go in developing crew/computer interfaces, but the problem is not as easy to solve as we might imagine. Full feature operating systems such as those found in modern computers take up a great deal of memory, absorb a lot of processing power in themselves and are prone to system crashes. The one thing that is absolutely mandatory for a spacecraft's computer system is reliability—there is no time to re-load software during an emergency. Memory also consumes power, which is limited when there are a large number of other systems to supply. The only significant way to improve the crew interface is by providing a much more graphically intensive display of information[7]—something that is memory- and processor-hungry in the extreme. Producing this sort of display reliably and with low power is a major challenge for future on-board computer systems.

Notes

[1] A gimbal is a kind of universal joint that allows free rotation within a range of angles. Old fashioned ships' compasses were often gimbal-mounted to enable them to remain horizontal while the ship rolled in heavy sea conditions.

[2] The history of the MIT Instrumentation Lab is interesting in itself. The department started as a small team of students and technicians working for Dr Charles Stark Draper (1901–1987). In 1973 it separated from MIT to become a non-profit research and development laboratory called The Charles Stark Draper laboratory, which is today still a leader in high-technology applications. Draper was a pioneer in the development of inertial navigation systems. His work started in World War II with the development of gun sights. During the 1950s further work on marine guidance was adapted for use in ballistic missiles. Since the laboratory's work on the Apollo guidance systems its largest contract has been Trident II. Further information can be found at http://www.draper.com/profile/docslab.htm and http://www.jsc.draper.com/. Currently the lab is under contract 'to the Johnson Space Center for guidance, navigation, and control support to the Space Shuttle program. This support covers the design of the powered flight guidance equations for ascent and on-orbit manoeuvres, the design and evaluation of the navigation system for all mission phases, and the design responsibility for the on-orbit flight control system for controlling on-orbit manoeuvres covering payload deployment and retrieval operations.'

[3] A logic gate is a circuit with a set of inputs and one output. The state of the output depends on the combination of signals presented to the input. It is a little like an electronically controlled gate that only opens when the correct code has been applied. The simplest sort of gate is the AND gate which will only turn ON its output if all the inputs are ON as well. An OR gate will turn ON if any one (or more) of its inputs are ON. A NOR (NOT OR) will only turn ON if all of its inputs are OFF.

[4] This is not a time clock—computer clocks provide the steady signal (rhythm) that is used to time sequences of operations that the processing unit has to carry out.

[5] ROM is Read Only Memory, i.e. memory that can be read, but not erased and over-written—it is like something that you can never forget. RAM is Random Access Memory, which can be read, erased and over-written.

[6] I've got one of these calculators. I bought it while I was at school. It was like buying a Rolls Royce of calculators. They are incredible machines!

[7] Those readers who are familiar with the film *2001* will remember the fascinating displays on the HAL 2000 computer that controlled every aspect of the ship's operations. Interestingly enough, it was while writing this section that I realized how bored the crew must have been on that (fictional) mission. The film shows them walking about, jogging, shadow boxing, fixing things that have gone wrong, lying on sun lamps—BUT NEVER DOING ANY SCIENCE...

Chapter 6

The lunar module

This chapter deals with the design of the lunar module ascent and descent stages, life on the surface, the activities of the astronauts, their spacesuits and the lunar rover.

6.1 Designing the first spacecraft

Without meaning to suggest that the Block II command module was anything other than the technological marvel that it undoubtedly was, I have always felt that it would have been more interesting to work on the design of the lunar module. After all, if we discount various space stations (on the grounds that they do not have means of propulsion), the lunar module is still the only proper manned spacecraft to have been designed and built. In saying this I am assuming that by 'proper' spacecraft we mean a vehicle designed to operate solely in the vacuum of space.

Free of the need to survive entry into an atmosphere, the design of the lunar module could evolve in a way dictated by expediency and the single-minded purposes of landing and providing a habitat on the Moon. The resulting design does not look aesthetically pleasing in any normal manner. Many have called the lunar module ugly. However, the eyes of an engineer can see the purity of purpose and intelligent design.

Dimensions:
Ascent stage height: 3.76 m
Descent stage height: 3.23 m
Diameter (diagonally across landing gear): 9.45 m
Earth launch weight: 14.5 tonnes
Pressurized cabin volume: 6.65 m^3

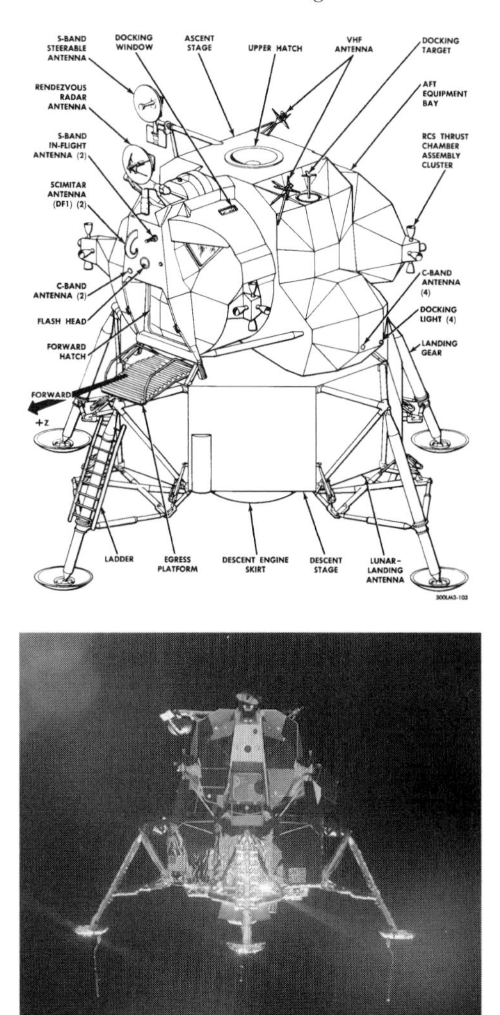

Figure 6.1 The Grumman-built lunar module. Top, the lunar module; bottom, the Apollo 11 lunar module viewed from the command module prior to landing.

6.2 The ascent stage

One key issue that was identified early in the development of the lunar module was visibility for the astronauts while landing on the Moon. An early design study by Grumman had a crew compartment reminiscent of a helicopter's with four large windows forming a bubble-like enclosure. Although this provided

Figure 6.2 This beautiful photograph of the Apollo 17 lunar module's ascent stage clearly shows how thin the skin of the vehicle was in places.

excellent visibility, the idea had to be dropped. The large glass area added too much weight as the glass would have to be quite thick to survive the pressure difference between the vacuum of space and the cabin environment. Not only that, but such a large glass area through which heat could be lost into space would have placed intolerable demands on the cabin's environmental unit.

Providing adequate visibility, especially downwards, continued to be a problem until a re-think in another area provided an unexpected solution.

Designing a seat to hold the astronauts in space suits, and yet give them adequate reach for the controls, proved to be rather difficult without taking up so much room in the cabin that the astronauts could not move about and get in and out of space suits. The solution to this was to take the radical step of removing the seats all together—an unheard-of departure in the context of a flying vehicle. In zero *g* and the low-gravity environment of the Moon, it was thought that the astronauts would be perfectly comfortable flying the lunar module standing up.

Once that decision had been taken, the visibility problem was easily solved as well. Two quite small triangular windows were provided that angled back into the cabin. While standing up the astronauts could get much closer to the windows and so had a much wider field of view—especially downwards. The size, shape and placing of these windows did much to fashion the unusual features of the ascent stage's 'face'. Furthermore, not having to worry about knee room for seated astronauts allowed the designers to shorten the length of the cabin, with consequent benefits in terms of weight saving.

Another design issue that went a long way to defining the shape of the ascent stage was the propellant storage for the ascent engine. It was always thought that the lunar module would essentially be a skin wrapped round the tanks and equipment needed. Early designs used four propellant tanks—two for the fuel and two for the oxidizer. This allowed the designers to balance the tanks on opposite sides of the ascent engine so that the centre of gravity of the stage was in line with the thrust direction of the engine.

However, Grumman became worried about the reliability of the design using four tanks and the associated plumbing, so the decision was made, with NASA approval, to do a costly re-design using only one tank for fuel and one for the oxidizer. The resulting design shaved 45 kg off the stage's mass, but cost $2 million. However, the upshot of this change was that the designers could no longer balance the tanks against each other. As the fuel weighed more than the oxidizer, the fuel tank had to be placed further away from the centre of gravity than the oxidizer tank, which gave the ascent stage its rather bulbous look on the right-hand side.

The cut-away drawing of the ascent stage's right half clearly shows the way in which the structure was sub-divided (figure 6.3). At the front was the cylindrical crew compartment, 2.35 m in diameter and 1.07 m deep. Looking at things from inside the lunar module facing the windows, the mission commander stood on the left-hand side with the lunar module pilot on the right. They were given armrests for use in the powered descent and cords with which they tethered themselves to the floor. It is worth emphasizing the size of the cabin that they were using. The metal decking provided about 1.5 m width of floor space, and midway between the windows the vertical clearance was 2 m.

A control panel ran at waist height across the front of the compartment. Mounted on this was a pistol grip attitude controller (rather like a modern computer game player's joystick) for each astronaut's right hand and a thrust control for the left hand. Between them on the commander's side was a start/stop button for backing up the main computer and on the lunar module pilot's side were the displays and data entry pad for the abort computer. Another waist-high panel was mounted between them and above the exit hatch. On this was the main entry and display panel for the guidance computer. As this extended about 30 cm into the cabin, the astronauts had to be careful not to hit it with their backpacks as they left the lunar module. The remaining instrumentation and controls were mounted above waist height on the forward wall of the cabin and on the bulkheads to either side.

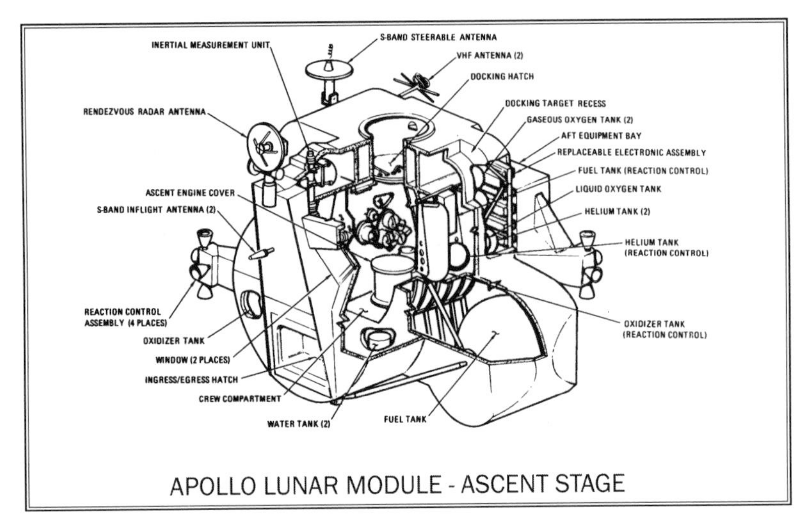

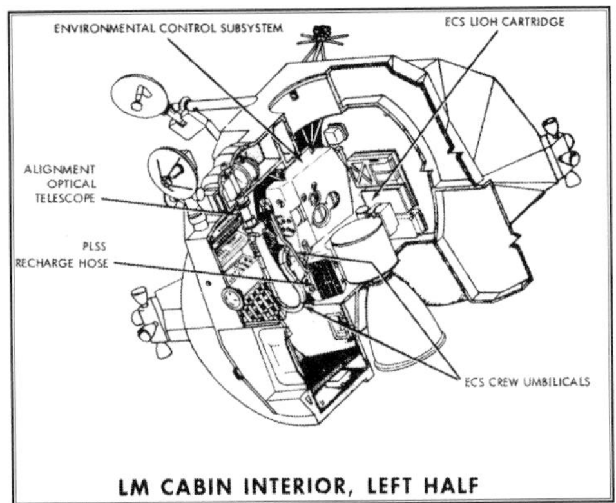

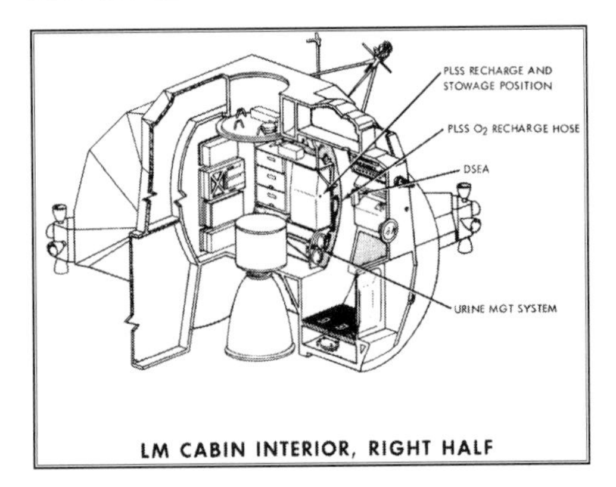

Figure 6.3 Cut-away drawings of the lunar module ascent stage.

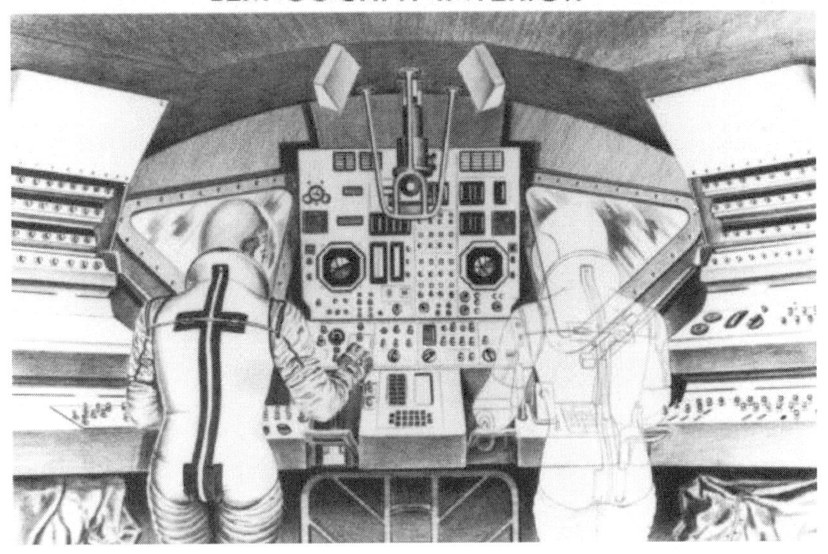

Figure 6.4 An artist's impression showing the inside of the ascent-stage crew compartment. The perspective has opened out the apparent space in the compartment. A thrust control can be seen under the commander's right hand as well as the computer panel between them and the telescope hanging down from the ceiling. The hatch is also visible under the control panel.

A small telescope for making star sightings while in orbit or on the surface hung from the ceiling over the computer panel with an eyepiece at eye level.

One interesting example of the astronaut's input into the configuration of the instrument panel involved the 'eight ball'[1]. This was a device for displaying an artificial horizon as a reference while flying the machine. Basically such instruments are freely mounted spheres. Newton's first law suggests that they remain in place unless a force acts on them, and being freely mounted the spheres tend to remain level as the cabin rotates (but as they are mounted into the panel they do move along with the spacecraft!). With a line drawn across the ball's diameter and a line etched into the glass cover of the instrument, the angle at which the spacecraft is flying is evident.

The two eight balls can be seen in the hexagonal mounts on the main panel in figure 6.4. Grumman had installed them on the lunar module assuming that the astronauts would wish to use them. NASA had them removed. The astronauts demanded that they be replaced, so they were. Next a NASA administrator asked why there was a development hold-up on the LM, and was told that there

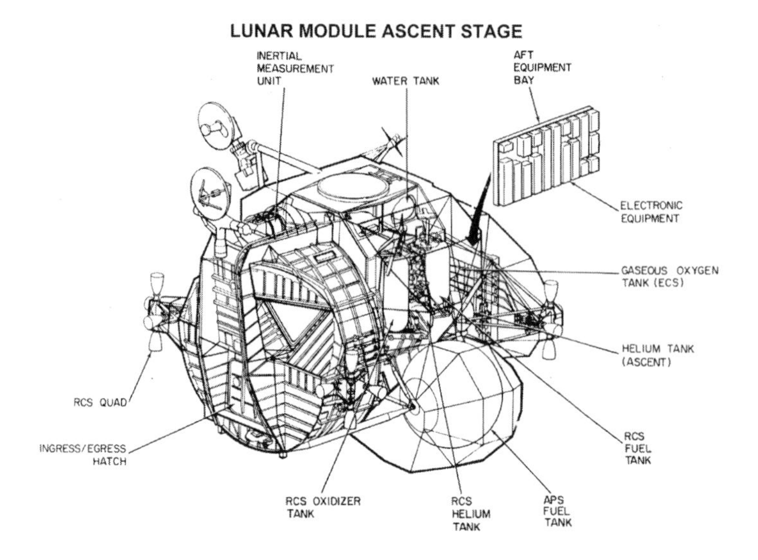

Figure 6.5 The construction of the lunar module ascent stage.

were problems such as the constant to-ing and fro-ing with the eight balls. He had them removed again. The astronauts found out and refused to fly the LM unless they were in place. They were put back, with a kit for easy removal!

Behind the crew compartment was the mid-section, marked by the docking tunnel at the top and the two propellant tanks either side at the bottom. The internal part of the mid-section was also pressurized and contained the guidance computer and the environmental control system.

Although the mid-section was a full 2.3 m deep at eye level, the space was curtailed by the docking tunnel above extending 40 cm into the cabin. The ascent engine cover below the docking tunnel came up to knee height. There was only 1 m of floor space in front of the ascent engine cover. The internal bulkheads were lined with equipment and storage compartments. On the commander's side were the stowage positions for the backpacks used on the lunar surface (Portable Life Support System—PLSS), as well as the 'urine management system'. On the opposite bulkhead was the Environmental Control System.

Outside the cabin in the mid-section region was the ascent engine fuel tank on the commander's side, with the oxidizer tank on the opposite side. Also dotted about between the mid-section outer skin and the internal bulkheads were LOX and LH_2 tanks as well as the propellant tanks for the lunar module's thrusters. The aft compartment of the ascent stage was used for equipment stowage.

Running over the whole ascent stage structure was a skin composed of two layers of Mylar, which provided some protection from small meteorites as well as thermal insulation, and a thin outer layer of aluminium.

Considerable debate went on regarding the hatches used by the crew to get into and out of the spacecraft.

The earliest designs assumed that there would be an upper hatch at which docking with the command module would take place when the LM was withdrawn from the Saturn third stage. However, when the ascent stage had to dock again with the command module having returned from the lunar surface, it was assumed that the front hatch would be used. The assumption seems to have been based on the need for the LM crew to see what was happening during the docking (reasonable enough as they would be flying the manoeuvre!), and this meant using the forward windows. The forward hatch would also double as the means by which the astronauts would climb out of the LM onto the lunar surface.

However, as with many aspects of the Apollo design, engineering a system to do two jobs forced compromises that could not be tolerated. In the case of the forward hatch it had to be reinforced to stand the stress of docking and be provided with a similar latching system to the upper hatch. These constraints made the hatch difficult to get out of in full spacesuits with lunar backpacks. Eventually it was felt easier to provide the lunar module with a small window in the top of the cabin, through which the commander could see during docking, and to dispense with the complex mechanics at the front hatch. After some trials with suited-up astronauts, the shape of the hatch was also altered to rectangular rather than the round shape that had been imposed by the docking idea—which made it easier to provide clearance for the backpacks.

Trying things out with the astronauts was a vital part of developing both the command and lunar modules. This was especially true of problems related to getting from the lunar module hatch down onto the surface. Grumman arranged for a pulley system (called a 'Peter Pan rig') to support most of an astronaut's weight in order to simulate the Moon's gravity. Using this they had various astronauts scramble over mock-ups of the lunar module, trying different ways of getting down to the surface. Even a knotted rope down which they could climb was tested. In the end a flat plate (porch) was placed in front of the hatch and a ladder attached to one of the landing legs.

The design of the ladder itself caused some debate. As nobody was quite sure how deep the fine dust on the surface was, they could not be certain to what extent the legs would sink. Consequently it was something of a guess as to how long the ladder should be!

Environmental Control System

One of the most important sub-systems on board the ascent stage was the Environmental Control System (ECS) which regulated the cabin's environment and supplied oxygen to the astronaut's suits while they were in the LM. The unit was situated on the bulkhead behind the lunar module pilot's station and the whole thing packed into a space about one metre high and half a metre thick.

Normal Earth atmosphere (at sea level) consists of nitrogen (78% by volume), oxygen (21% by volume), water vapour and traces of other gases such as CO_2. Of this cocktail the oxygen is by far the most important component for us when we breathe. Working on pure oxygen enabled the cabin pressure to be considerably reduced—down to 33 kPa, which is roughly the same as the pressure exerted by just the oxygen content of the atmosphere at sea level. Consequently the LM's walls did not have to be designed to be quite so strong (saving weight), and the ECS did not have to be complex enough to deal with more than one gas—which would have added weight and reduced reliability. Half jokingly, the astronauts were rather worried about the possibility of kicking a hole in the LM's skin!

As the astronauts breathed they inhaled pure oxygen, with a little water vapour, from the cabin and exhaled an amount of CO_2. If this was not filtered out of the cabin supply then the CO_2 would slowly build up to dangerous proportions. As part of the ECS system, lithium hydroxide canisters were used to filter out the CO_2. Fresh oxygen was supplied to maintain the cabin pressure.

During the Apollo 13 mission the CO_2 levels built up dangerously in the LM. The ECS system was designed to deal with the respiration of two astronauts. With the command module systems crippled, all three crew members were having to survive in the LM and consequently the lithium hydroxide filters were becoming saturated. Unfortunately the filters used in the command module were the wrong shape to fit into the lunar module. A team in mission control came up with a way of adapting the filters by using hose from the space suits, the covers from the flight plan (now somewhat out of date), masses of tape and various other parts that could be scrounged from the spacecraft. Tense moments were spent building this kludge following instructions radioed from the ground, but the system worked.

The ECS was designed to maintain the cabin pressure at 24 kPa for at least 2 minutes should a 13 cm hole be punched in the cabin wall. Fortunately this facility was never required during the Apollo programme. The only times that a moderate risk of puncture was foreseeable was during landing, ascent from the surface (in both cases debris could be kicked up from the ground), or during docking. During those manoeuvres the crew wore their suits, although

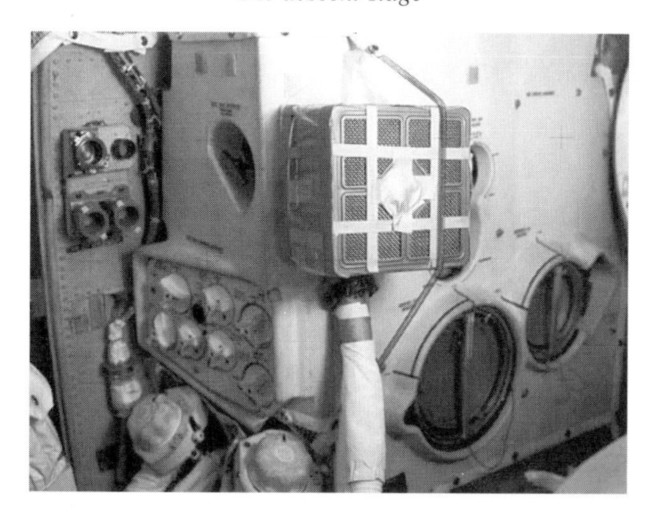

Figure 6.6 The do-it-yourself fix to the lunar module's CO_2 filtering system used during the Apollo 13 mission.

not inflated so that they were able to move about more freely. Two minutes was time enough for them to inflate the suit to survive depressurization (the problem was not loss of oxygen to breathe so much as loss of pressure to act against the body's internal fluids).

Depressurization of the ascent stage prior to going outside (or throwing out the trash) could be done using one of two dump valves. One of these was outside the craft in case the ascent stage accidentally re-pressurized while the crew were on the surface—this would make the hatch impossible to open against the pressure difference. Pulling on the external valve could reduce the cabin pressure from 34 kPa to 0.5 kPa, low enough for the hatch to open, in about 180 seconds.

6.3 The descent stage

The descent stage served three purposes. Firstly, it acted as a mount for the descent engine which, as I commented in chapter 3, was possibly the greatest technological breakthrough achieved by Apollo. Secondly, the descent stage had to act as a stable platform from which the ascent stage could be launched back into orbit. Finally, it acted as a storage bin for various pieces of equipment—experiments and geological tools, latterly the lunar rover, replacement batteries, spare lithium hydroxide canisters for use in the LM's air recycling systems and a two-day supply of food. Anything that could be stowed down there was, as it meant more room for the astronauts in the ascent stage and less mass to carry back into orbit (and so less propellant for the ascent engine).

Figure 6.7 The underside of the lunar module descent stage. The round objects are the propellant tanks. One can also see the top mounts for the landing legs.

Weight reduction was a critical aspect in the design of both the ascent and descent stages of the LM. However, due to the demands imposed by the functions that the descent stage had to carry out, the scope for paring weight was rather less. Not only was the stage required to be rigid enough to support the lift off of the ascent stage, it had to support the landing legs and part of the shock of impact as well. Furthermore, the descent engine (with all its technology) was larger than the ascent engine and it required a greater propellant load. Consequently the descent stage accounted for two thirds of the total LM weight at launch.

The basic design of the descent stage was cross shaped (as can be seen in figure 6.7) and sub-divided into square sections. The descent engine occupied the centre section, with the propellant tanks arranged symmetrically round the outside to balance the weight distribution. The flat sides of the outer squares were used as the attachment points for the landing legs and the remaining triangular sections (formed by joining the edges of the squares together) were used as equipment stowage bays.

The Space Technology Laboratories Inc. (STL) designed descent engine used a hypergolic propellant mix of hydrazine and nitrogen tetroxide. The propellant flow rates were controllable either manually or via the on-board computer system. As with the service propulsion system, the propellants were force-fed to the engine by pressurized helium from a storage tank.

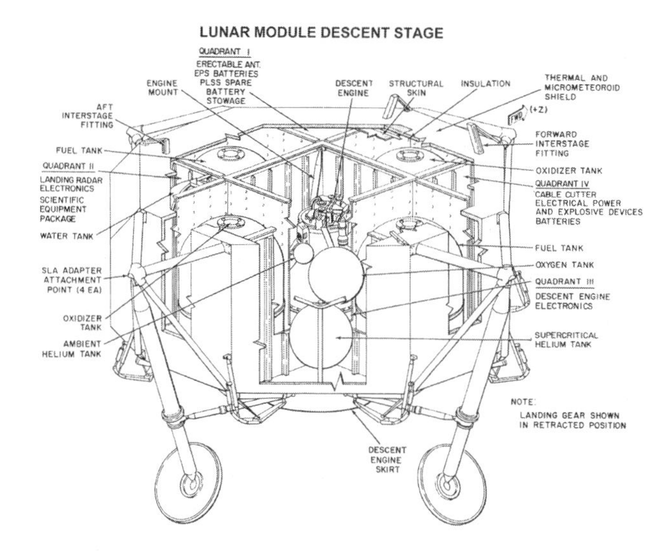

Figure 6.8 The internal design of the descent stage.

The landing gear was designed to support the shock of impact on the lunar surface and to prevent the LM from tipping over—providing it landed on a slope of less than 6°. Each leg consisted of a primary strut with an inner and outer cylinder in a piston-like arrangement. The inner cylinder contained a cartridge of honeycombed aluminium that compressed on impact, thus absorbing the forces. The primary strut was designed to deform by up to 81 cm on landing. Running the length of the forward strut was a ladder for the astronauts to descend from the top porch to the surface.

Attached to the end of each landing gear was an aluminium honeycomb footpad with a diameter of 94 cm. This large diameter was used to minimize the pressure that the weight of the LM exerted on the lunar surface.

Protruding from each footpad (bar the front one) was the *lunar surface sensing probe* reaching 1.52 m below the base of the pad. These electromechanical devices caused an indicator light to illuminate on the LM's console (the contact light) when they came into contact with the surface. Once the contact light came on, the astronauts were supposed to shut down the descent stage allowing the LM to fall the remaining distance to the Moon. The length of the probe was determined as a compromise—too long and the LM would hit the surface at a speed that would cause the forces in the landing struts to exceed their design levels; too short and the LM stage could be damaged by the heat of the descent engine reflecting from the surface as well as the exhaust reflecting back or kicking up stones.

Powered descent[2]

Once the lunar module had separated from the command and service module combination it flew in front so that the command module pilot could make a visual check of the spacecraft (for one thing he confirmed that the landing legs had correctly extended to clear the engine bell of the descent engine). Then the command module used its thrusters to perform a radial burn, placing it on an orbit that tracked 4 km above the lunar module's path during the descent. The command module came into line with the lunar module half an orbit later on the lunar far side.

At this moment the lunar module's descent engine was fired to slow the craft down and move it into an elliptical orbit, the firing point becoming the apocynthion, with a pericynthion of 15 km at a point 14° up-range of the landing site. This burn had to be performed on the far side of the Moon as the landing site was to be on the near side (for obvious communication reasons) and it takes half an orbit to move from apocynthion to pericynthion.

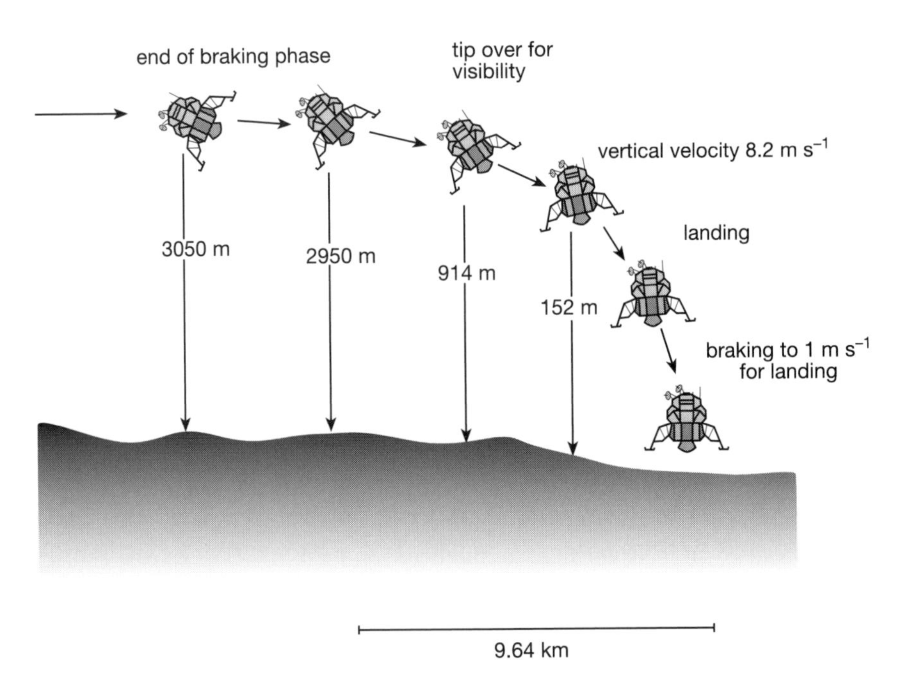

Figure 6.9 The final phase of powered descent. Although this graphic suggests that the lunar module was coming in to land sideways, the craft was in fact flying on its back for the braking phase and pitched forwards to allow the crew to see the surface during the final stage.

Figure 6.10 The 'flying bedstead' used in powered-descent training.

Nearly three hours after the separation the lunar module reached pericynthion and the descent engine was fired again to brake the craft out of this orbit. From this point to just before the landing the computer flew the spacecraft. At the start of the burn the LM was 482 km from the landing point and travelling at 1.7 km s^{-1}. The first phase of the burn was designed to slow the forward velocity essentially to zero. At the end of this braking phase the computer used the thrusters to tip the lunar module forwards so that the crew could see the surface.[3]

Having made a visual assessment of the surface features, looking for landmarks and a safe landing spot, the crew were able to manually take over flying. With the correct throttle setting the descent engine could counter the gravitational pull of the Moon, allowing the lunar module to hover. Thrusters could then be employed to skim the craft around to find a landing spot. Finally the craft entered a nearly vertical descent, slowing the vertical velocity from 8.2 m s^{-1} to 1 m s^{-1} prior to landing. Once the probes on the LM's footpads made contact with the surface the engine was shut off and the craft settled onto the Moon.

During the landing, the mission commander was responsible for flying the spacecraft while the lunar module pilot backed him up by cross checking his actions and relaying instrument readings to him (such as altitude, rate of fall and horizontal velocity). During the Apollo 14 flight, another problem arose after the intermittent abort signal mentioned in intermission 3 had been solved. The landing protocols called for the lunar module's radar to lock on to its own reflections from the surface so that it could provide the computer with accurate height readings. If this had not happened by the time the lunar module had descended to an altitude of 4 km, the flight plan called for the lunar module to be pitched upright and the abort sequence to start. Unfortunately Shepard's radar stubbornly refused to see its own reflections. Once again the lunar landing was being threatened. The problem was eventually fixed by switching the radar's power off and then on again so that it re-set itself. A lock was established shortly after this had been done.

Training for the mission-critical process of powered descent was done by a combination of mission simulations inside full mock-ups of the LM and flying a most ungainly craft called the 'flying bedstead'. This was a rather unstable single-man training vehicle that was designed to simulate the LM's characteristics. Several of the astronauts had rather hairy moments training in this device.

6.4 Space suits

Walking on the Moon is not like walking in the park. The lack of atmosphere presents more of a problem than simply not being able to breathe. The human body has evolved in an environment which subjects it to an external pressure of some 100 kPa (atmospheric pressure). To cope with this the internal body fluids exert a pressure outward on the skin. An astronaut on the Moon without suitable protection would explode due to the pressure difference between the inside of his body and the vacuum of space. Furthermore, the lack of atmosphere means that sunlight is not scattered. The direct light from the Sun is far brighter on the Moon than on Earth. Consequently the heating effects are much greater. Over 50% of the energy reaching the Earth from the Sun is either reflected or absorbed on its way through the atmosphere. On the Moon, this does not happen. Standing in bright sunlight on the lunar surface would be extremely hot. Also, the atmosphere provides protection from the ultra violet light produced by the Sun. Unprotected skin would burn rapidly on the Moon. Finally, without scattered light or atmosphere to conduct and convect energy, shadow areas on the Moon are extremely cold. The astronaut has to be protected from extreme temperature variations. The role of a space suit is then to:

- contain a pressurized environment to support the body;
- provide breathing oxygen;
- provide protection against thermal losses and excess heating;
- provide some protection against radiation;
- provide some protection against micrometeorites, scrapes against the lunar module or rocks on the surface that might otherwise puncture the suit.

Two versions of the space suit were designed for Apollo. An ‘intravehicular pressure garment assembly’ was worn by the command module pilot and ‘extravehicular pressure garment assemblies’ were worn by the commander and the lunar module pilot. These were identical except for the outer layers on the extravehicular units, which were designed to protect the wearer from micrometeorites and had extra thermal protection. The following descriptions are of the full Moon-walking suit.

Each astronaut had his own individual set of three suits ‘tailored’ for him[4]. One was for use in training (inevitably this was the one subject to the most wear), another was the main suit for use on the mission, and the third was a back-up in case a fault or rip developed in the main suit during the preparations for launch.

Working from the skin outwards, the astronaut first wore an intricate set of long johns knitted from nylon–spandex through which a series of plastic pipes carried cooling water to extract excess heat from the body. During the normal in-flight routine the crew wore Teflon cloth flight coveralls over the long johns.

Prior to launch and while on the Moon they put on the complete space suit system. Over the long johns came the multiple-layer pressure suit. The innermost layer was designed for comfort and was made from lightweight nylon. Over this was a neoprene-coated nylon pressure bladder skinned with a nylon layer to give it strength. This bladder inflated when air was pumped into the suit at a pressure of 26 kPa. Together the bladder and the nylon layer provided the chief resistance to pressure difference.

Working inside an inflated suit was rather like trying to move inside a rigid balloon. In order to achieve greater flexibility, rubber joints were provided at the knees, waist, elbow, shoulders and other areas. The joints were formed like bellows and reinforced with built-in restraint cables.

Next came a sequence of five aluminiumized Mylar layers (for thermal insulation) interspersed with four layers of Dacron. After this were two further layers of Kapton and beta marquisette (providing extra thermal insulation), a layer of Teflon-coated filament beta cloth (to give protection against tearing and wear) and finally a white Teflon cloth layer. The final Teflon layers were designed to be flame proof.

The standard pressure suit came with shoes, but for walking on the Moon a lunar overshoe was made for each astronaut. The outside of the shoe was constructed from woven fabric and metal with a ribbed silicon rubber tread. The tongue was of Teflon-coated glass-fibre cloth. Inside layers were made from the same material alternating in 25 layers with Kapton film to give insulation.

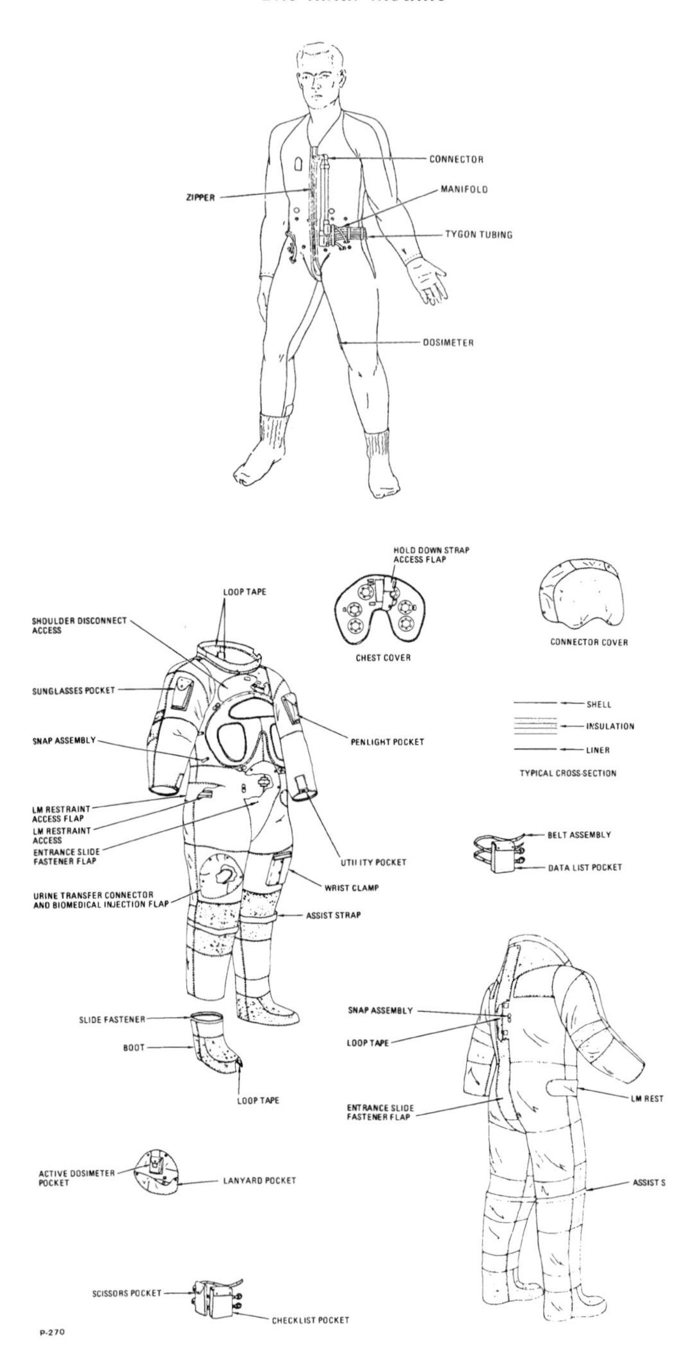

Figure 6.11 The Apollo space suit. Top, the cooling underwear; bottom, the main space suit.

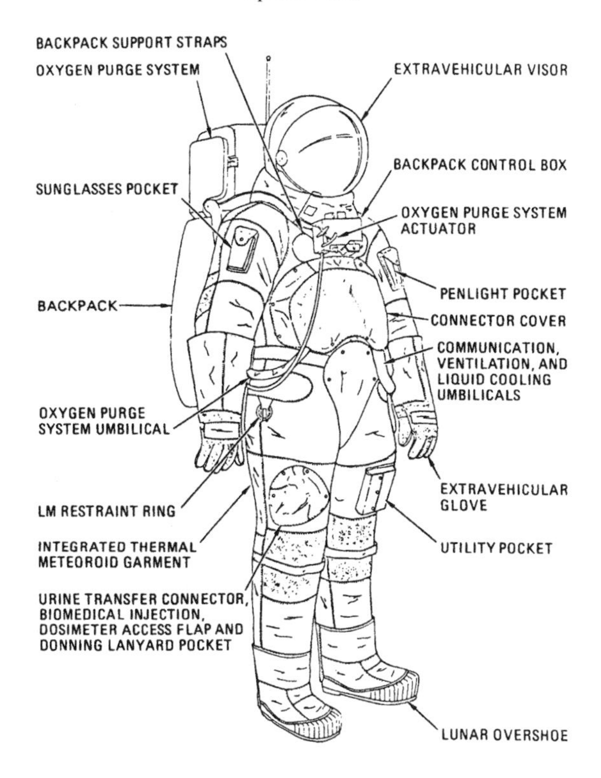

Figure 6.11 (*Continued*) The main space suit with the backpack, lunar overshoe and lunar visor.

Gloves were a problem for designers as they had to be flexible enough to enable the crew members to handle controls and tools. Two types of glove were used. Black rubber ones for launch were made from a mould taken of the astronaut's hand. The lunar surface gloves had silicon rubber fingertips which allowed the astronauts some degree of 'feel'. Both types of glove were connected to the suit by a clip and locking ring.

The final component of the suit was the helmet. For launch there was a 'fish bowl' helmet made from transparent polycarbonate with a small cushion behind the head. On the Moon a hood was pulled over the helmet, with a gold plated visor to protect the eyes from the Sun.

During pre-flight checks the crew waited in the suiting-up room and oxygen was provided to them in their suits by a small portable unit. They carried this unit to the transport van when they were called to board the rocket. Once safely strapped into the command module, their suits were connected to the capsule's main environmental control unit.

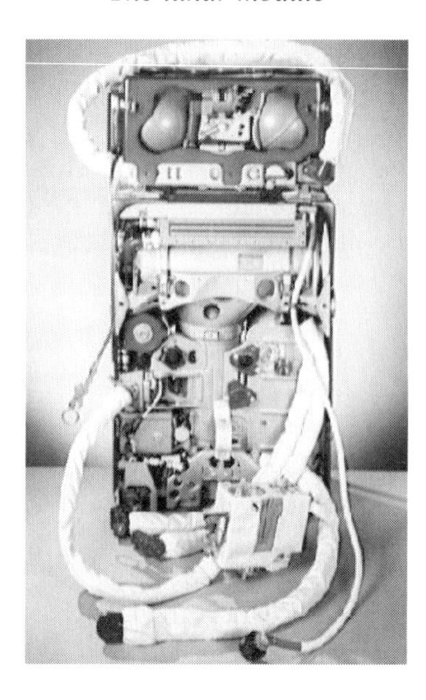

Figure 6.12 An Apollo backpack.

Walking on the Moon required a much greater degree of freedom, and that both hands be free. Consequently while on the surface the astronauts wore a sophisticated portable life-support unit on their backs.

Backpacks

Containing the variety of support systems that an astronaut needed while walking on the Moon in a lightweight (353 N on Earth) unit was a masterpiece of engineering. Not only did the backpack carry breathing oxygen (0.8 kg), and the fans to circulate it round the suit, but also the lithium hydroxide filters needed to extract CO_2, cooling water and pump, a radio, a battery and the equipment to re-charge the suit from the LM supplies.

The cooling-water pipes in the astronaut's long johns were linked to a connector on the outside of the suit and from there via a hose to the backpack. Pumps in the backpack circulated water round the pipes without loosing significant amounts. A second supply, called the *feedwater*, was passed at a controllable rate through a heat exchange system. Here it was placed in thermal contact with water from the cooling circuit and conducted energy away, reducing the temperature of this water for passing back into the suit. The hot feedwater was allowed to evaporate into the vacuum. The backpack held about 5 kg of feedwater, which was sufficient to provide cooling for about 8 hours of quite strenuous activity.

In case of an emergency, each astronaut carried an 18 kg Oxygen Purge System (OPS) on top of the backpack. In the event of a backpack failure, a suit puncture[5], or running out of the main oxygen supply, this system could provide a back-up supply. The same system was intended to supply the suits should the LM be incapable of forming a hard dock with the CSM—in which case the astronauts would have to make a space walk. If the cooling system in the suit was still working the OPS could be set to supply oxygen at a rate of 1.8 kg h^{-1}, giving about one hour of supply. With no cooling system working on the suit the OPS had to be set to a high flow rate which only gave about 30 minutes of breathing oxygen. On later missions equipped with a lunar rover (see the next section), the Moon walk plans ensured that the crew were never more than one hour's direct driving time away from the LM—in case an emergency return was required. In such circumstances the two suits could be linked together so that cooling water from one backpack could supply the other one, allowing the slow flow setting to be used.

One of the last things that the astronauts did before leaving the Moon was to depressurize the LM, open the door and throw out the trash—along with the now useless backpacks.

6.5 The lunar rover

A couple of years ago the Audi car company ran an advert for their A8 aluminium-bodied saloon car. It featured one of the engineers who had worked on the lunar roving vehicle making a comparison between the two aluminium vehicles. At one level the comparison is obviously a marketing stunt; at another there are some interesting features to compare. Both cars used aluminium extensively for low weight, and both had four-wheel drives for stability.

The Apollo lunar rover was designed and built by the engineers at Boeing, commissioned by the Marshall Space Flight Center. The rover was powered by two 36 V batteries, although as usual with redundant design either battery alone was sufficient to operate all the on-board systems.

Having a mass of 210 kg, the vehicle could manage a range of 65 km in total and was capable of 10–12 km h^{-1} on level ground. The range used in practice was restricted to a 9.5 km radius which gave a reasonable 'walk-back' distance to the LM should the rover be immobilized.

If you take into account the various tasks that had to be performed to ready the rover and shut it down after use, the astronauts did not gain much time in using it to travel over the surface compared to being on foot. The advantage lay in

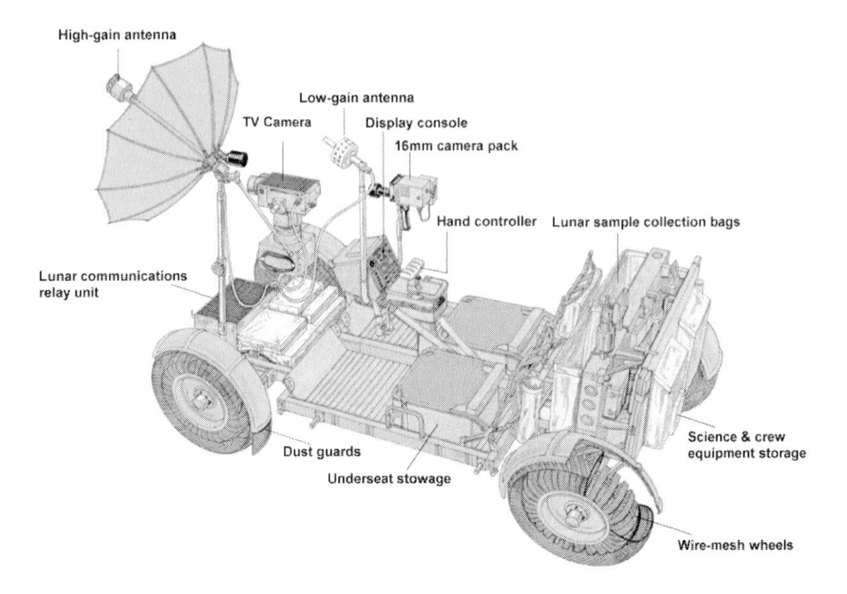

Figure 6.13 The Apollo 15 lunar rover.

their using less oxygen, being less tired and being able to carry more samples and equipment about.

The combination of low lunar gravity and a surface covered in fine dust provided an interesting challenge for the design of the rover's suspension. Each of the wheels was made from a strong zinc-coated piano-wire woven mesh to which were riveted titanium treads forming a chevron-shaped series of 'tyre' grooves (they can just be seen in figure 6.14).

While being driven the rover kicked up a great deal of dust—hence the need for a 'mud guard' over the top of each wheel. I remember watching the astronauts driving about the surface, testing out one of the rovers, throwing dust into the air[6] which would fall back to the ground in a curiously slow motion due to the low lunar gravity. The rover was designed to be robust and it certainly came under a degree of abuse while on the surface. The wheels were independently driven by electric motors (about 1/4 horsepower each) and there was also an independent steering system for both front and back wheels.

Possibly the most fascinating thing about the rover, however, was its on-board navigation system. Astronauts discovered that finding their way about on the lunar surface was often harder then they expected. The combination of crystal-

Figure 6.14 The Apollo 15 lunar rover in use on the surface of the Moon.

clear viewing (no atmosphere to blur distant objects) and the total uniformity of colour made it very hard to judge distances. It was quite easy to spot the LM on the surface as it was highly reflective and a totally different colour, as well as standing 7 m high—so there was never any real danger of being lost. However, a great deal of planning went into the EVAs from the geological point of view and the scientists on Earth often wanted samples taken from precise points. The rover's navigation system helped with this.

Operating the navigation system was relatively straightforward. Before a trip started the system was re-set at a point near to the LM. As the astronauts set off the system counted rotations of the rover's wheels to determine distance and used that in combination with an on-board gyro which provided directional information. In this way it could pinpoint the rover's location in distance and direction from the starting point to an accuracy of about 100 m. However, the system was useless as far as directing the crew to a specific location was concerned until the precise landing point of the LM had been fixed. The lunar rover was used on the last three Apollo landings and, in each case, there was some initial uncertainty regarding the landing spot. Eventually the astronauts managed to unambiguously recognize some features (craters, etc) that enabled those on the ground to fix a location from which they could direct the EVAs.

The rover carried a high-powered aerial to relay communications to Earth via the lunar module. There was also a remote-control TV camera and a movie

Figure 6.15 The take off of an ascent stage as filmed from the remote control camera on a rover.

camera mounted on the chassis. The TV camera was motorized and could be directed by the controllers on Earth. Aside from providing stunning pictures of the astronauts' activities while on EVA, the camera also panned about, observing the lunar panorama. The most memorable view relayed to Earth, however, was the launch of the lunar module ascent stage (figure 6.15). The rover was parked a safe distance away and the camera was panned upwards as the ascent stage made its way into orbit.

6.6 The ascent to orbit

With their stay on the lunar surface completed, the two astronauts would stow their equipment and rock samples and fire up the lunar module's ascent engine to return to orbit. The powered ascent was divided into two phases—vertical ascent and orbital insertion. Vertical ascent was a flight straight upwards from the lunar surface to ensure that the LM had cleared the terrain before the craft pitched over so that the ascent engine was accelerating the LM's horizontal motion up to orbital speed. Pitchover started once the LM's vertical ascent rate had reached 15 m s^{-1}, which was some 10 seconds after the burn had started and at an altitude of 76 m. The ascent engine continued to burn for a total of about seven minutes achieving a ΔV of 1.8 km s^{-1}. Once the powered ascent had been completed the LM was settled into an orbit which was 16.7 km by 81.6 km (relative to the surface). Having done that the ascent engine had completed its task for the mission. All the remaining manoeuvres would be carried out by using the reaction control system's thrusters on the ascent stage.

Once the fine details of the orbit achieved were available to them, the crew would use the LM's main computer to calculate the parameters for the final manoeuvres leading up to rendezvous with the CM in its circular 111 km orbit. Once the LM had reached apolune[7], the RCS thrusters would be fired to achieve a ΔV of about +15 m s^{-1} which would move the LM's orbit into one that was about 28 km below that of the CM (84.3 km by 81.9 km). Being lower than the CM's orbit, the LM would be moving faster and so would catch up on the higher spacecraft at a rate of about 0.072° per minute. Other burns of the RCS thrusters took place once the craft were nearer to each other to slow the catch up rate and finally to bring the craft into docking. The whole sequence from the start of the first RCS burn took some three and a half hours.

Notes

[1] The eight ball is an important ball in billiards.

[2] The exact details of how powered descent took place varied from flight to flight—this account is based on the landing of Apollo 11.

[3] At the start of the powered descent the LM was horizontal and face down so that the crew could see the surface for an initial landmark check. The crew was also able to estimate their height from the rate at which the surface features were moving past (the computer gave their linear velocity, they could estimate their angular velocity out of the window, so the radius of their path could be estimated). After this initial phase the LM would be low enough for the radar to lock onto the surface so the ship was turned so that

the crew were flying on their backs, feet down. This meant that when the ship pitched forwards during the last phase of descent they could see the landing site in front of them as they were approaching it.

[4] The modern EVA suit used on the space shuttle is more universal. Only the gloves are individually tailored.

[5] Of course, this would not repair the hole. The oxygen would still leak out, but the system could go some way to maintaining the pressure while the astronaut returned to the LM.

[6] Just an expression—there is no air on the Moon, of course!

[7] Apolune is the point on the orbit where the spacecraft is furthest from the centre of the Moon.

Intermission 4

The three 'ings'

Eating

> *Balanced meals for five have been packed in man/day over-wraps, and items similar to those in the daily menus have been packed in a sort of snack pantry. The snack pantry permits the crew to locate easily a food item in a smorgasbord mode without having to 'rob' a regular meal somewhere down deep in a storage box.*

This extraordinary quote, which makes the Apollo flights sound like a cross between a Sunday school picnic and a military exercise, comes from NASA's Apollo 11 press kit. Food for the astronauts was carefully planned to provide a balanced intake over the period of the flight. The restrictions of zero *g* dictated that the food selection was packaged as freeze dried and rehydratable, wet packed, or potted to be eaten with a spoon.

In the command module there was a dispenser for normal drinking water, as well as two taps that produced measured amounts of water warmed by the excess heat from the fuel cells (to either 68 °C or 13 °C). Water was injected into a food package which the astronaut then 'kneaded' for several minutes to ensure a good mixture. Finally he cut off the top of the package and squeezed the contents into his mouth. Once the astronaut was done he placed some germicidal pills into the packet, to prevent fermentation of the remains, rolled up the bags and threw them into a waste compartment.

Stored along with the food were toothbrushes and a two-ounce tube of toothpaste for each crew member. Supplies of wet-wipes and tissues were also stowed in various places about the spacecraft.

Table 4.1 Neil Armstrong's menu on the Apollo 11 flight. NB: Day 1 consisted of meals B and C only.

Day 1,5	Day 2	Day 3	Day 4
Meal A			
Peaches Bacon squares Strawberry cubes Grape drink Orange drink	Fruit cocktail Sausage patties Cinnamon toasted bread cubes Cocoa Grapefruit drink	Peaches Bacon squares Apricot cereal cubes Grape drink Orange drink	Canadian bacon and apple sauce Sugar-coated corn flakes Peanut cubes Orange–grapefruit drink
Meal B			
Beef and potatoes Butterscotch pudding Brownies Grape punch	Frankfurters Apple sauce Chocolate pudding Orange–grapefruit drink	Cream of chicken soup Turkey and gravy Cheese cracker cubes Pineapple–grapefruit drink	Shrimp cocktail Ham and potatoes Fruit cocktail Date fruitcake Grapefruit drink
Meal C			
Salmon salad Chicken and rice Sugar cookie cubes Cocoa Pineapple–grapefruit drink	Spaghetti with meat sauce Pork and scalloped potatoes Pineapple fruitcake Grape punch	Tuna salad Chicken stew Butterscotch pudding Cocoa Grapefruit drink	Beef stew Coconut cubes Banana pudding Grape punch

Sleeping

Sleeping in zero *g* was a novel experience. Most of the astronauts reported problems in sleeping due to not being used to having no support from a pillow or covers weighing down on the body. Sleeping crewmen would also tend to float about the cabin, which is why sleeping bags which could be tethered in place were provided. The situation was slightly easier on the Moon with some gravity to help. However, astronauts in the LM faced their own problems when it came to sleeping arrangements. On the early missions the crew were forced to spend the whole time on the Moon in their spacesuits. The inflexibility of these bulky suits made it difficult to get comfortable. By Apollo 15 a suit re-design made them easier to remove in the LM's cabin and from that flight on the crew were able to strip to their long johns in order to sleep. On the Apollo 11 flight the commander (Neil Armstrong) and the lunar module pilot (Buzz Aldrin) slept perched on the ascent engine cover and curled up on the floor respectively. Later missions carried beta cloth hammocks, one of which could be strung across the cabin and the other strung above the first running along the cabin's length.

In any case, after several days in the command module the unfamiliar sounds of the LM's systems and the excitement of being where they were made it difficult to rest.

Excreting

Of all the natural functions that the crew were required to carry out on board Apollo the equipment developed to help with 'waste management' was the most primitive and unpleasant.

Urine was dealt with by a hose with a condom-like fitting at one end. The hose lead via a valve to the vacuum of space. The principle was to attach the hose, open the valve and urinate into space where the liquid would freeze into crystals that shimmered in the sunset. Apparently the sight was quite spectacular. In practice, however, the system was not that easy to operate. Aside from the psychological problems involved with connecting part of your anatomy to the vacuum of space, opening the valve created a pressure difference that tended to draw you into the hose. Shutting the valve also had the tendency to trap your anatomy in the hose as well.

If this aspect of space flight was not the most convenient then the solid waste management was even worse. That consisted of a top-hat-shaped bag with a sticky rim. The idea was to attach the bag in the proper position and to proceed ensuring that the material remained in the bag. Not the easiest of jobs in the zero *g* environment of a spacecraft. The final delightful task was to place a germicidal pill into the bag and to knead the contents to ensure a good mix. The bags were then stored for analysis back on earth (*nice work if you can get it!*).

The system devised for the space shuttle is much more sophisticated and can be used by both male and female astronauts. It is more akin to the toilet found on a modern jet airliner, with the addition of spring-loaded thigh restraints!

Chapter 7

The shuttle and its followers

In this chapter we will look at the design and operation of NASA's space shuttle fleet. We will also consider what happened in the tragic explosion of Challenger and consider the future development of re-usable space systems.

7.1 The space shuttle

The space shuttle represents a radically different approach to the process of carrying astronauts and equipment into Earth orbit.

Some people argue that NASA should have continued to develop the rocket planes that the United States Air Force was flying in the late 1950s (such as the X-15) as a means of getting into space, rather than committing itself to non-re-usable booster rockets. While such designs would have had a re-usable capability, NASA took the view at the time that their development would not be fast enough to enable project Mercury to beat the Russians into space. With the race to the Moon won, NASA was able to return to the idea of a system that could shuttle back and forth into space. With a large number of the system's elements being used for flight after flight, the launch costs should come tumbling down, enabling commercial exploitation of space. NASA started to divert funds from Apollo to the development of the space shuttle during the Nixon administration.

The space shuttle is a beautiful, complex and innovative machine. Its design attempts to combine the aerodynamic characteristics of an aeroplane that can glide, land and be re-used with the rocket technology that can provide the thrust needed to get into orbit. Yet it was designed in the 1970s, and first flew in 1981; the technology is rooted in those days.

Arguably the shuttle is due to be replaced by something cleverer.

Figure 7.1 The first shuttle sits on its mobile launch pad. Note that on the early flights the external tank was painted to match the orbiter. The large white structure to the left-hand side is the payload changeout room. This swings round to mate with the orbiter and provide a weather-tight seal so that the payload bay doors can be opened. The vertical gantry on the very far left is the support for the structure. At the bottom of this gantry is the cab in which the driver sits when the structure is driven round on a track in an arc to mate with the orbiter.

There are three components to the space shuttle system. The *orbiter* is the plane-like part that rides into orbit on the back of a large orange-coloured cylinder—the *external tank*. Inside the external tank are two separate vessels containing LOX and LH_2. These liquids are pumped to the main shuttle engines situated at the back of the orbiter. Strapped to either side of the external tank are the *solid rocket boosters* (SRBs). These long cylinders contain moulded solid propellant and provide 71% of the thrust at launch. Without the SRBs the shuttle could not leave the ground. They are jettisoned after 2 minutes of flying, by which time the mass of the craft has reduced sufficiently (due to the consumption of propellant in the external tank) for the main engines to carry on accelerating it into orbit. The SRBs then parachute back into the ocean where they are recovered and used again. Each SRB casing is designed to last 20 launches. The orbiters are designed for a minimum of 100 launches.

Once empty the external tank is jettisoned and will burn up in the atmosphere. It is the only part of the shuttle system that cannot be re-used[1].

The orbiter can carry a crew of up to seven people[2] and remain in orbit for several days (the current record is over 17 days). It can deploy satellites into orbit, capture and return to Earth faulty satellites, transport scientific experiments, repair satellites *in situ* (such as the Hubble Space Telescope), and latterly has started the assembly of the ISS.

On return to Earth the orbiter orients itself so that the underside is facing down and slightly forwards. It then uses its supplementary engines (the orbital manoeuvring system engines—which use hypergolic propellant) to slow down and trigger a gravity-assisted descent into the atmosphere. The actual path taken by the orbiter is complex and designed to minimize the effect of air resistance on the craft. It is largely flown automatically by the on-board computer systems, but the pilots are able to take over in case of an emergency. Once through the re-entry phase the orbiter glides at supersonic speeds to its landing strip—it has no engines that are capable of guiding it through the atmosphere. The orbiter touches down on a runway like a conventional aeroplane (except somewhat faster—between 341 and 364 km h^{-1}!) and brakes to a halt with the aid of a parachute.

7.2 Shuttle components and construction

The orbiter

There are several major sections to the construction of the orbiter.

Starting from the front of the craft, the *forward fuselage* (which extends from the nose back to the start of the payload bay doors) has upper and lower sections that fit together like a clam shell. The forward RCS (reaction control system) module contains thrusters for manoeuvring in space. Further thrusters are mounted in the blisters on either side of the tail fin (vertical stabilizer).

Inside the forward fuselage is the pressurized vessel of the *crew compartment* (rather like the inside of a vacuum flask). This is a welded aluminium structure that sits inside the forward fuselage and is connected to it at four attachment points. With so few attachment points the thermal conduction through to the compartment from the outside during re-entry is minimized, as well as the loss of heat while in space.

Access to the crew compartment is via the main entry hatch on the left-hand side of the orbiter (viewed from above). On the mobile launcher the white room at the end of the swing arm connects with this hatch. In an emergency the hatch can also be blown open. The second way in and out of the crew

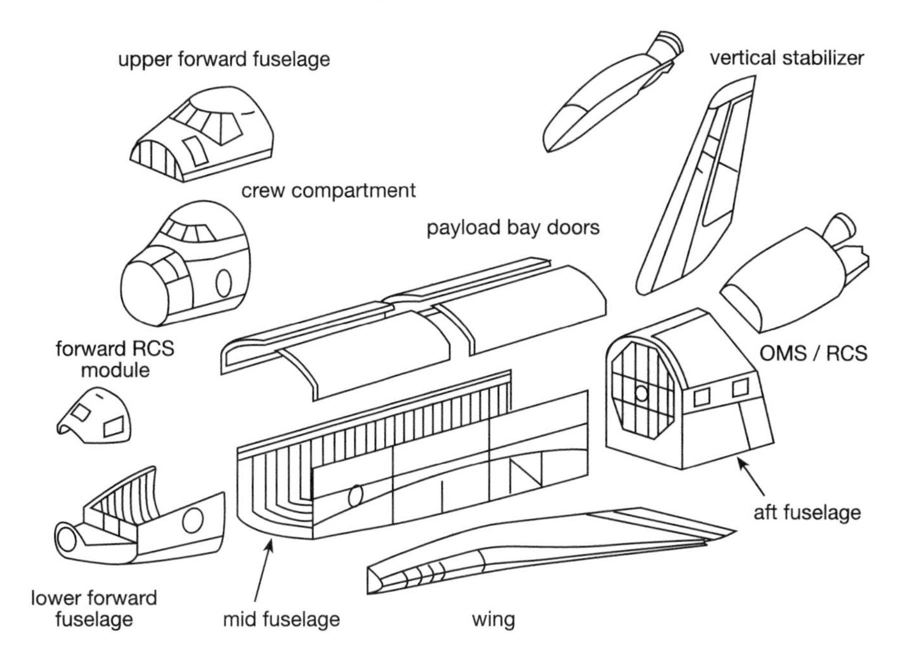

Figure 7.2 The main components of the orbiter.

compartment is via an airlock that passes through the pressure vessel and the rear bulkhead of the forward fuselage. This gives access to the huge payload bay. On some shuttle missions a further pressurized module—Spacelab, Spacehab or the docking adapter for connecting with the Mir space station—is mounted in the payload bay. The crew can float into the module via a tunnel linked to the airlock.

The crew compartment is divided into two decks. The top deck, which is effectively the blister that can be seen on the upper surface of the orbiter from the windows back to the payload bay doors, is the flight deck where the commander and shuttle pilot sit during launch. In front of them is a control panel that looks similar to that found inside a modern airliner. It is, however, considerably more complicated. Behind the seats for the commander and pilot is a work area. This has two viewing windows that look back along the spine of the orbiter and another pair that look vertically upwards. Here the deployment of satellites and other payloads from the bay can be controlled, as well as the remote manipulator arm. The shuttle can also be manoeuvred from this area. A further two seats can be mounted on the flight deck during launch (later stowed away). Part way down on the left-hand side is an opening in the floor with a ladder extending down into the mid-deck—the ladder is, of course, superfluous in zero g.

The mid-deck can house a further three or four seats during launch, depending on which orbiter is in use, and is the main living area for the crew. It also contains three equipment bays. Two are at the front of the deck, taking up the whole of the forward bulkhead wall, and one is at the back to the side of the airlock. Aside from containing equipment, the lockers in these bays can also hold the crew's personal items. The 'waste management system' is next to the main hatch. This is a sophisticated unit that can cater for both male and female clients (unlike that provided for Apollo).

Depending on the space taken up by equipment on the deck, a galley and a washing unit can be installed as well. If no galley is installed then a suitcase-sized food warmer is provided. Sometimes ice cream and frozen steaks can be provided if a refrigerator is being used for biomedical experiments (it is important to mark the contents clearly!). Sleeping arrangements vary depending on the space taken up by equipment stowage. The crew can sleep in their seats, in sleeping bags that can be tethered to the wall or, if there is room, a bunk-bed system can be mounted on the mid-deck.

The crew compartment contains a nearly normal mixture of 80% nitrogen and 20% oxygen at sea-level pressure. The air is filtered and recycled. During launch the crew use pressure suits very similar those worn by fighter pilots. During an EVA the shuttle space suits are worn. For day-to-day mission operations the crew can use jeans and T-shirts or any clothing that they would wear on Earth.

The *mid-fuselage* is 18 m long and is the main load-carrying structure of the orbiter. It transmits the force of the engines from the rear fuselage to the rest of the craft and contains the structure that links the wings together. At the top of the mid-fuselage is the *payload bay* with its large hinged doors constructed from a graphite epoxy frame with honeycomb panels. Once in orbit these doors open (irrespective of whether the mission has a payload to deliver) as their undersides contain the panels that allow the orbiter to radiate excess heat produced by equipment into space. For most of the mission the orbiter will circle the Earth with the payload bay open and facing down towards the ground. This ensures that much of the direct sunlight falls on the underside, where the most heat-resistant surfaces are placed.

After the mid-fuselage comes the *rear fuselage*, onto which are mounted the shuttle main engines (burning LOX and LH_2 from the external tank) and the two smaller OMS engines. The vertical tail fin, with its two flaps that open out as air brakes on landing, is also mounted on this section.

The wings are constructed of aluminium alloy with a honeycomb cover.

Remote manipulator arm

The first of the remote manipulator arms flew on the Columbia orbiter in 1981. The arm was a gift from Canada to NASA and subsequently NASA has ordered more of them, with a total export value of $400 million. The arm is used for a variety of purposes. Primarily it is designed to grab satellites and guide them into the payload bay (alternatively, it can be used to deploy objects from that bay). During the mission to repair the Hubble Space Telescope the arm linked to the telescope, pulled it into the payload bay, carried the astronauts (functioning as a mobile work platform) and finally delivered the refurbished telescope back into orbit.

Other more unconventional uses have included knocking away a block of ice from a clogged waste-water vent that might have endangered the shuttle upon re-entry, pushing a faulty antenna into place, and successfully activating a satellite (using a swatter made from briefing covers) that failed to go into proper orbit.

Figure 7.3 The Hubble Space Telescope (the large cylinder in the centre of the picture) was repaired and refurbished by the crew of STS-61 in December 1993. The crew member visible in the photograph is being moved on the end of the shuttle's remote manipulator arm.

Although on Earth the arm cannot lift its own weight, in orbit it can accurately manoeuvre payloads of 30 000 kg. It comprises an upper and lower arm boom with various joints and an end effector where various components can be linked. The arm is guided from a station at the back of the flight deck using translational and rotational hand controllers to direct its movement.

Thermal protection

Spacecraft before the space shuttle orbiter survived the heating effect of atmospheric friction during re-entry by the use of ablative heat shields. These are one-use shields that char and flake apart as they are heated, carrying energy away from the craft. As the orbiters were intended to be re-used at least 100 times, a new way of protecting them from heat needed to be developed.

It is not just the heat of re-entry that must be catered for. In orbit the shuttle can be warmed to 121 °C in direct sunlight and cooled to −156 °C in the shade. The shuttle uses a variety of materials to provide thermal protection and new systems are continually being developed to improve their properties and reduce costs.

The orbiter's nose and under its chin are, together with the leading edge of its wings, the hottest areas during re-entry. About 20 minutes before touchdown the temperature on these surfaces climbs as high as 1600 °C. These areas are protected with a carbon composite which is moulded in one piece for the nose and in several sections for the wings. Some areas are further insulated with quartz blankets and metal foils. The composite is constructed from a graphite cloth which is soaked in a special resin. Layers of the cloth are then laminated together and the whole structure baked in an oven, which converts the resin into carbon. A mixture of aluminium, silicon and silicon carbide is then coated over the top.

Much of the rest of the orbiter's surface is insulated by a combination of 24 300 tiles and 2300 flexible insulation blankets. The tiles account for 70% of the remaining surface.

There are two types of tile used. The high-temperature re-usable surface insulation (HRSI) tiles (black ones) cover the underside of the orbiter, the region round the flight deck windows and various other parts (most of the black bits other than the wing edges and the nose are the black tiles). The other tiles, low-temperature re-usable surface insulation (LRSI) (white ones), were originally used extensively on the areas of the orbiter that do not have to survive extreme heating.

Figure 7.4 This close-up of Discovery's nose (taken just before it was lifted by crane to be mated with the rest of the shuttle in the vehicle assembly building) shows some of the different thermal protection materials used: carbon composite on the tip and upper part of the nose, black tiles on the underside of the nose, and white tiles on the upper nose around the windows.

Both tiles are made from fibres of pure white silica which has been refined from sand. These fibres are mixed with high-purity de-ionized water, along with other chemicals, and poured into a mould. Any excess liquid is squeezed out of the mixture. The blocks are then dried out using the largest microwave oven in America at the Sunnyvale plant of Lockheed Space Operations Co. Then they are sintered in a 1290 °C oven, which fuses the fibres without melting them.

All of the tiles on the orbiter are different shapes. Each one has a curvature on its underside that exactly matches that of the orbiter's body at the place where the tile is to be applied. The blocks are first cut with saws and then milled to the correct shape by diamond-tipped machines. The manufacture of each tile is fully documented and an individual tile can be traced back to the batch of sand that it was made from.

The final step is to coat the tiles with borosilicate glass into which various chemicals are added to give the tiles their different colours and heat-rejection properties. Surface heat dissipates so quickly that a tile can be held by its corners a few seconds after removal from a 1200 °C oven, while the centre of the tile still glows red.

The tiles are delicate and each one has to be inspected, repaired or replaced after every mission.

Most of the LRSI tiles have now been replaced by flexible insulation blankets (FIBs). These are made from a waterproofed, quilted fabric with silica felt between two layers of glass cloth sewn together with silica thread. These blankets have better durability, and cost less to make and install than the tiles. They are used on the upper sides of the orbiter's fuselage, sections of the payload-bay doors, most of the tail fin, parts of the upper wings and around the observation windows.

After each flight it is necessary to replace the waterproofing agent, which has burnt away during re-entry. This is done by injecting new agent into each tile through existing small holes. The blankets are also treated, using a needle gun.

The external tank

Once loaded with propellant, the external tank becomes the heaviest space shuttle component as well as being the largest. Internally it is divided into a LOX tank at the top, an inter-tank structure in the middle and an LH_2 tank at the bottom. It is 47 m long and has a maximum diameter of 8.4 m. The tank's mass when empty is 30 tonnes.

The external tank is connected to the orbiter at three places. The forward attachment point is under the shuttle's nose about halfway between the tip and the flight deck windows. The two aft attachment points are either side of the tank and connect to the orbiter on the mid-fuselage just behind the end of the wings. In the same region, umbilical lines carry propellant, gases, electrical signals and electrical power to and from the orbiter. The control signals for the solid rocket boosters are also routed through these umbilicals.

Computer systems

The shuttle orbiter is the most computer-intensive spacecraft yet built. It includes a full digital fly-by-wire avionics system. In conventional aircraft, controls such as rudder pedals, stick and hydraulic actuators of control surfaces (ailerons, flaps, etc) are operated by mechanical linkages and cables. Fly-by-wire, which is becoming increasingly common in aircraft of both military and commercial designs, replaces such mechanical links with sensors and electronic signals to operate electrical actuators. Similar technology is even finding its way into road cars—for example, some BMWs use variable resistors to sense the depression of the accelerator pedal and send this information to the engine's computer, rather than relying on a cable to open a throttle valve. However, the orbiter was one of the first flying machines to use this means of controlling aspects of its flying performance.

Adding to the complexity of the computer systems is the switch to more autonomous control of various aspects of the ship's performance that used to be governed by ground-based systems—for example fuel levels and life support. Finally, as the crew has to interact with the computer systems on a far more regular basis than was the case on any previous spacecraft, a significant amount of software and hardware is devoted to the display of information and the commanding of the computers.

There are five computers on board the orbiter, operating in a redundant manner. Off-the-shelf designs from IBM were chosen rather than specialist machines built specifically for the application, as had been the case with Gemini and Apollo. However, the input and output circuits and memory components were purpose-made. Another change from Apollo was the switch to using a high-level programming language[3]. Previous systems were programmed using machine-level language that was expensive and difficult to de-bug. A specialist programming language HAL/S was developed for the shuttle computers.[4]

The earliest shuttle computers were IBM AP-101 central processing units which could run 480 000 instructions per second. They shared access to the core memory (104K on the first flight) and the re-writeable mass memory units. The systems have been updated since flights resumed after the Challenger disaster.

7.3 Assembling a shuttle

The process of preparing a space shuttle for another flight begins with the retrieval of the spent solid rocket boosters from the ocean. Two dedicated ships (*Liberty Star* and *Freedom Star*) sail to the points where the boosters splash down and tow them back to shore. They are then inspected, cleaned, and parts of them are shipped back to the propellant company in Utah to be refilled with solid propellant.

The SRBs are designed to split into several segments. The top three segments contain (among other things) the control electronics, the parachutes and the igniter that sends a 45 m jet of flame down the length of the SRB to set fire to the propellant on lift off.

Next come the four segments that contain the solid propellant mixture. By mass the mixture proportions are 69.6% ammonium perchlorate oxidizer, 16%

Figure 7.5 Towing a recovered solid rocket booster back to shore.

aluminium fuel, 0.4% iron oxide catalyst, 12.04% polymer binder to hold the mixture together, and 1.96% epoxy curing agent.

The propellant has to be moulded into the booster in comparatively short segments that are stacked together again as the mixture has to cure for 5 hours, during which time it is important that the heavier components of the mixture do not settle to the bottom. This is easier to prevent in shorter segments. Matched pairs of boosters are produced for a given flight by making sure that two SRBs are loaded with propellant from the same batch of mixture. This is important—if one booster had more thrust than the other, the craft might not leave the ground vertically!

The first propellant segment is moulded into an eleven-point star with the points facing inward. The remaining segments are internally moulded into a truncated cone. This allows the boosters to provide maximum thrust at ignition and then a decreasing thrust (down by about 1/3) 50 seconds after ignition—which corresponds to the time when the shuttle starts to feel the maximum pressure from air resistance. The lowest segment with fuel packed into it also contains the convergent/divergent nozzle, which can gimbal by up to 7° in order to steer the shuttle during an ascent. After this segment comes a skirt assembly which takes the weight of the entire shuttle when it is resting on the mobile launcher. It also contains four small motors to separate the boosters from the shuttle about two minutes into the flight. There are four more such motors at the top of the booster stack.

Great care has to be taken in joining the segments together as the weight of the packed fuel mixture can distort the casing, making it hard to mate the parts again. Most of the jointing is done at the factory, but some of the joints (the field

Figure 7.6 Stacking segments of the solid rocket boosters.

joints) are made and sealed 'in the field' inside the vehicle assembly building (VAB). A completed SRB is over 45 m tall, 371 cm in diameter and weighs 590 tonnes at launch.

The segments arrive at the Kennedy Space Center by rail (it is too dangerous to fly them) and are then used to start the process of assembling a complete space shuttle. First the lowest segments of the SRBs are stacked on support posts atop the mobile launch platform. Then the remaining segments are stacked up on top until a pair of completed SRBs are standing on the platform. Assembling the boosters is time-consuming and difficult work. The connecting bolts and joints must be rigid enough for the entire stack to function as a complete unit during the stress of launch.

As each external tank is only used once, a new one must be supplied from the manufacturers for every launch. These arrive at Kennedy from New Orleans by barge as they are too large to fit in a truck or aeroplane. Inside the VAB a gigantic crane lifts the external tank above the mobile launcher and lowers it between the two SRBs that are already standing there. The crane supports the mass of the empty tank while engineers bolt it in place between the boosters. This involves several days of painstaking work to ensure accurate alignment.

While all this has been going on an orbiter has been prepared in the orbiter processing facility (OPF), a building constructed since the end of the Apollo programme next to the VAB. The OPF consists of three almost identical 31 m tall buildings containing hangars and maintenance workshops. Three orbiters can be serviced in the OPF at one time.

The first shuttle flights ended with a landing on one of the extra-long runways at Edwards Air Force Base (California). The shuttle was then loaded atop a converted 747 jumbo jet and flown back to Kennedy. Now that a great deal of landing experience has been gained, shuttle missions end with a landing on the specially constructed runway at Kennedy. However, the jumbo is still available should a landing take place elsewhere (such as during a launch abort or if bad weather prevents a landing at Kennedy).

In any case, once the shuttle is back at Kennedy it is wheeled into the OPF where an intensive inspection and service routine lasting between four and six weeks is carried out.

The craft is jacked up off its landing wheels and surrounded by a complex system of scaffolding that allows access to every part of the orbiter's body. Each one of the heat-resistant tiles is inspected by hand and damaged (or missing!) ones replaced. The main engines and fuel pumps are removed and sent for refurbishment. Sometimes the same engines are then put back, or alternatively ones from a different orbiter or even totally new ones are used instead. All the electrical circuits and computer systems are thoroughly checked, along with the guidance systems and the life-support components. The inside is also thoroughly cleaned!

Once the checking is over the orbiter is lowered onto its landing gear and wheeled across to the VAB. Meanwhile, the mated external tank and SRBs on the mobile launch platform have been surrounded by another complex of scaffolding to allow technicians access to as much of the system as possible. This means that the orbiter has to be lifted off its landing gear into a vertical position and then winched over the top of the scaffold and down the other side to mate with the other components. This involves lifting the orbiter over 100 m from the floor of the VAB.

The crane supports the orbiter while it is bolted to the external tank. Once this has been done the whole shuttle assembly is supported on the solid rocket boosters.

Figure 7.7 Landing at Edwards Air Force Base (plenty of clear space!) and then returning the orbiter to Kennedy.

The next stage is to drive the crawler vehicle into the VAB and under the mobile launcher. Jacks are then used to lift the launch pad clear of its supports within the VAB and it drives out on the crawler as Apollo did in the past.

The crawler, mobile launchers and launch pads are all modified versions of the original components used for Apollo.

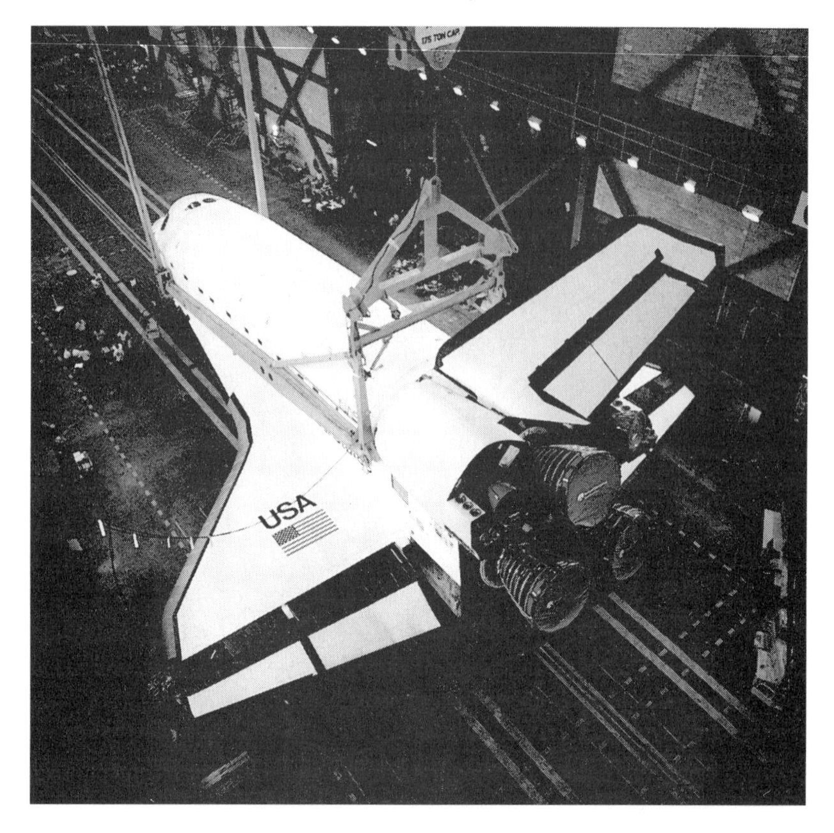

Figure 7.8 Lifting an orbiter into a vertical position to be attached to the other shuttle components in the VAB. The three large disks are the covers for the main engines. On either side of the top engine can be seen the covered manoeuvring engines for use in orbit.

Installing the payloads

The orbiter is capable of carrying a large variety of payloads in its bay. Large structures that require a direct connection to the shuttle for power or for life support are loaded on board while the orbiter is horizontal in the OPF. Examples of such payloads would be Spacelab, Spacehab or perhaps an experiment pallet for use while in orbit. Other payloads, which are generally satellites to be deployed in orbit, are loaded into the payload bay while the shuttle is at the launch pad. The payload changeout room referred to in the caption for figure 7.1 is instrumental in installing payloads while the shuttle is on the pad. First the payload is loaded into a vertical canister and then hoisted into the room while it is in its open position for launch. Then the structure closes on the orbiter to provide a clean room in which the payload can be finally prepared and loaded on

Figure 7.9 Discovery is lowered towards a stacked external tank and SRB combination in the VAB.

board the shuttle. The payload changeout room is large enough for the orbiter's payload-bay doors to be opened out. Having the facility to install payloads while the shuttle is on the pad gives extra flexibility in loading payloads onto whichever shuttle is ready for launch.

7.4 Launching a shuttle

Like the Saturn V before it, the space shuttle requires many hours of preparation before a safe launch can take place. The countdown sequence generally starts at T−48 hours, which is actually about 72 hours from launch as NASA factors in several pauses (holds) in the countdown to allow for delays and faults that might develop. Some of the more interesting specific events in the countdown sequence are as follows.

- T−6 hours, when a planned hold takes place to allow engineers to finish getting the external tank ready for the propellant to be loaded on board. At this point the launch director must make sure that the weather forecast

Figure 7.10 Discovery is lowered into place. Note the connecting pipes that carry propellant from the external tanks to the main engines at the bottom of the external tank.

shows favourable conditions for the projected launch time[5]. There is no point in pumping the propellant into the tanks if the weather is not going to last. Once the hold ends the first job is to cool the pipes that will carry the cryogenic propellant into the tanks on the shuttle. Without this step a great deal of the super-cold liquid would boil into a useless gas while it was being transferred from the storage tanks at the pad.

- The shuttle's tanks have been filled by the time that the countdown reaches the T−3 hours mark. By this stage a Saturn V would have had ice forming on the outside of its upper stages. However, the shuttle's external tank is coated with insulating foam that prevents water vapour in the air freezing on its surface. It also helps to minimize the amount of propellant that boils inside the tank. The reason for doing this is that any ice falling from the side of the tank as launch takes place might strike the orbiter damaging one or more of the delicate heat-resistant tiles that cover the surface. On early missions the insulating foam was painted white so that it matched the rest of the shuttle. However, as this cost several thousand dollars it was soon decided that the aesthetics of the shuttle could cope with a rusty-looking external tank.

Figure 7.11 Roll out of an assembled shuttle from the VAB.

- There is another planned hold at the T−3 hours mark. During this time an engineering crew inspect the external surfaces of the tank for ice. If any has formed the hold is extended to allow it to melt. At the same time the crew is woken up. In the days prior to launch the crew steadily alter their sleeping pattern so that no matter what the actual time of day the projected launch takes place early in their morning. The crew undergo a physical examination, eat breakfast and don their pressure suits for transfer to the orbiter.
- The crew leave their quarters at T−2 hours 55 minutes. Having taken a lift to the top of the service tower next to the shuttle, they cross the gantry to the white room—a clean room mated to the external hatch of the orbiter. Waiting for them there are a team of technicians who will help them strap into the couches for the launch. At this point, with the orbiter standing vertically, the crew are lying on their backs in heavy pressure suits for the remaining countdown.
- Over the next couple of hours the team in the launch control building next to the VAB keep watchful eyes on the weather, pressure in the propellant tanks and fuel lines, as well as numerous other systems. Inside the orbiter the commander and pilot work on checking systems and aligning the guidance controls.

- Another built-in hold comes into effect at T−20 minutes. This short delay allows the launch director to give last-minute instructions to his team in the control room.
- The final pre-planned hold comes into effect at T−9 minutes, when the launch director asks each of his system managers in turn for final launch approval. Once this 'go' is given the countdown is under the control of the automatic ground launch sequencer and things start to happen in increasingly rapid sequence.
- T−7 minutes 30 seconds—the crew access gantry with the white room swings out of the way.
- T−6 minutes—the auxiliary power units on the orbiter power up to provide the power for engine gimballing.
- T−4 minutes 30 seconds—the orbiter switches to running from electricity generated from its own fuel cells rather than that provided by ground-based systems.
- From T−4 minutes to T−3 minutes the orbiter's engines are automatically gimballed to ensure that they are functioning properly before launch. At the end of this test the engines are moved to their launch positions.
- T−1 minute 57 seconds—the cap on top of the external tank which allowed boiling LOX to vent away is retracted.
- T−31 seconds—the orbiter's on-board computers take over the countdown.
- T−11 seconds—water is pumped into the trench below the launch pad and also to the sprinklers that spray water across the top of the launch platform. This is another modification to the Apollo system. The incredible sound pressures built up by the SRBs on launch could reflect off the metal surface of the launcher and damage the orbiter. The free-flowing water across the surface serves to dampen these sound waves and minimize the problem. The first orbiter launched lost 16 tiles and had 148 damaged, mainly due to the pressure waves generated at launch. Subsequently the water suppression systems were modified.
- T−8 seconds—at this time those that are watching the launch on TV will see a myriad of sparks playing about beneath the main engine nozzles. As the engines ignite within seconds of this, it is often thought that these sparks are responsible for triggering the engines. In fact what is happening is that the flow of LH_2 to the engines starts some seconds before the LOX is released so that there is a hydrogen-rich environment in the thrust chamber. Some of this hydrogen leaks out of the engine bell and would form a highly explosive 'pool' under the engines but for a device on the pad that activates to burn off the hydrogen and prevent it gathering. This is what causes the sparks.
- T−6.6 seconds—first main engine ignites.
- T−6.48 seconds—second main engine ignites.

- T−6.36 seconds—final main engine ignites. The three engine starts are staggered to gradually increase the load on the shuttle rather than forcing it to take the full strain from the beginning. The force of the main engines causes the top of the shuttle to tip by 1–2 metres. The super-hot exhaust of the main engines enters the trench beneath the mobile launcher, flash-heating the water there into steam which billows around the base of the pad in the moments before launch.
- Having checked that all the main engines are running at 90% or more of their nominal thrust, the computer signals the SRBs to ignite at T−0 seconds. Once this has happened the launch cannot be halted. The shuttle is heading into the air come what may. At any moment up to the final ignition of the SRBs the main engines can be shut down and the launch scrubbed. The exhaust from the SRBs is a highly visible bright jet of flame, unlike that from the main engines, which is invisible—being super-hot water vapour.
- Once the shuttle has climbed above the gantries and scaffolding of the mobile launcher, flight control is handed over from the Kennedy Space Center to the Johnson Space Center in Houston (as it was in the days of Apollo as well). Shortly after the shuttle 'clears to tower' it starts a 'roll and pitch programme' that turns it over so that the orbiter is flying upside down and at an angle to the vertical. The roll gives the pilots a clear view of the horizon (which would otherwise be obscured by the external tank) so that in the case of an abort that required them to glide either back to Kennedy or to some other landing strip they are able to orient themselves immediately. The purpose of the pitch is to ensure that the force of gravity acting downwards on the shuttle as it climbs steadily pulls the flight path more and more nearly parallel to the Earth's surface.
- At about 50 seconds into the flight the speed of the craft has built up to 1190 km per hour and the altitude has reached 9.1 km. Round about now the frictional forces on the shuttle build up to a maximum. Earlier than this the shuttle was moving through denser air, but more slowly. Later than this it will be moving more quickly, but the air will be much less dense at that altitude. The on-board computer throttles back the main engines (and the SRBs will be reducing thrust at this time due to the design of the internal propellant moulding) to ease the transition through this maximum dynamic stress region. Shortly after this the engines are throttled back up again.
- Just after the two-minute mark the SRBs are jettisoned. Although they have not yet burned out, they are reducing thrust. They continue to climb to something like 67 km altitude and finally parachute back to the sea about 227 km from the launch site.
- Four minutes and thirty seconds into the flight the shuttle's speed, altitude and distance from the launch pad make a return to Kennedy impossible. If an abort happens now the craft will have to glide to another landing site[6].

- By six minutes after launch the shuttle is moving at 20 000 km h^{-1} in the horizontal direction. Its flight path has brought it to the point where the orbiter is completely upside down and travelling parallel to the Earth's surface. However, it is not yet moving quickly enough to maintain an orbit. The engines continue to burn and the shuttle starts a slow 2 minute long descent back to Earth. In doing so it picks up more velocity due to the engines and is aided by the pull of gravity—losing between 4 and 6 km of altitude in the process. During this time the main engines gradually throttle back. The external tank is now mostly empty so if the engines continued at maximum thrust the much reduced mass of the shuttle would mean that the acceleration would build beyond the rate which the structure could survive. Acceleration is kept to a maximum of $3g$.
- The main engines are turned off 8 minutes and 30 seconds after launch. The shuttle is now in an elliptical orbit with a perigee of 65 km and an apogee of 296 km above the Earth's surface. Twenty seconds later the explosive bolts holding the external tank in place fire and it falls back to Earth, burning up in the upper atmosphere.
- The orbit that the crew now find themselves in is too low to be maintained. Friction with the upper atmosphere at perigee would eventually slow the craft and cause it to re-enter. For this reason, as it reaches apogee (about 50 minutes after the main engines cut out) the orbital manoeuvring system engines fire, producing a ΔV of +240 km h^{-1} and circularizing the orbit to 296 km altitude[7]. It might seem easier to burn the main engines longer and to inject the orbiter directly into this orbit, but that would require a larger external tank to carry the extra propellant and in turn that would be harder to accelerate off the ground. The technique used turns out to be more efficient in terms of propellant mass used. Not only that, there are also safety concerns—jettisoning the external tank at the height of the final orbit makes it much more difficult to control where it will burn up. It would not be good PR to have large pieces of external tank hitting the ground in populated areas.

7.5 Challenger

The morning of 28 January 1986 was cold. NASA had never launched a space shuttle in such low temperatures before (2 °C, a full 15 °C lower than the previous coldest launch). NASA will probably never launch a shuttle like this again.

At 11:38 Eastern Standard Time the shuttle lifted off on the tenth flight of the Challenger orbiter. Seventy three seconds later a catastrophic explosion ripped the external tank apart, threw the SRBs away to steer drunkenly across the sky,

Figure 7.12 The morning of the Challenger launch was unseasonably cold. Left, ice forming under the launch pad; right, ice in the vicinity of the SRB (visible behind the scaffold along with the orbiter's wing).

and blasted the orbiter into pieces. Most people viewing the scene assumed that the seven crew members had been killed instantly in the huge explosion. Some saw the SRBs descending in the distance on their parachutes and thought that the crew had escaped. In fact, analysis of the film shot of the launch clearly showed that the flight and mid-decks of the orbiter had been thrown clear of the explosion and fell 14 km to the sea amid a rain of debris. Challenger had been travelling at about 650 m s^{-1} when the explosion occurred. Examination of the recovered debris showed that the crew compartments had been virtually undamaged by the explosion—the damage that was present had been caused by the impact with the water. It is possible that some of the crew members survived up to the point of impact. However, it is very likely that none of them was conscious, due to the loss of oxygen when the decks were thrown clear of the explosion that punctured the cabin.

The events of that morning burned into the American consciousness like nothing since the Apollo 1 fire. The TV coverage had been extensive and many people saw the terrible events live. The reason for the close coverage was that one of the crew members, Sharon Christa McAuliffe, was the first civilian in space. A schoolteacher. Her class was looking forward to talking to her while she was in orbit. In a manner this symbolized the American dream—work hard enough and you can do anything, even fly in space.

Figure 7.13 Evidence of an unsealed joint: smoke emitted from the right-hand SRB.

President Reagan appointed former Secretary of State William Rogers to chair a committee to investigate the accident. Among the people selected to take part were Neil Armstrong and the famous American theoretical physicist Richard Feynman. All shuttle flights were postponed until the investigation discovered what had gone so badly wrong. The next shuttle mission did not fly until 29 September 1988, which was the only launch that year.

The key events in the launch sequence leading up to the disaster were as follows.

- Photographs taken by the cameras arranged about the launch pad showed that 0.678 seconds after the shuttle lifted off, puffs of grey smoke emerged near to one of the *field joints* at the bottom of the right-hand SRB (see figure 7.13). A total of nine such puffs were observed up to 2.5 seconds after lift off. The smoke emerged with a frequency of about four times per second, which roughly matched the frequency with which loads were being placed on the SRB structure by the launch, and hence the way in which the joint was flexing. The pictures showed each puff of smoke spreading

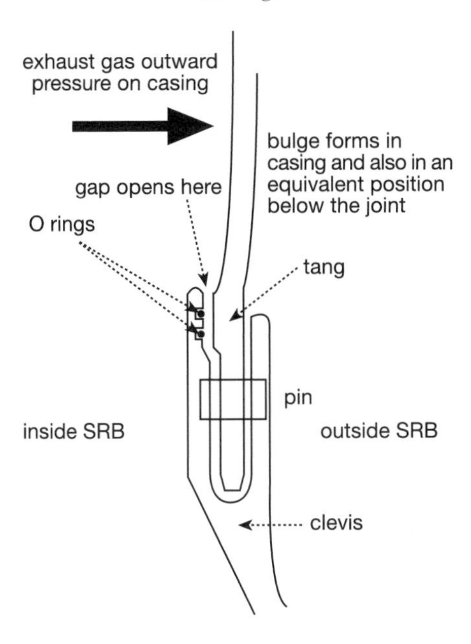

Figure 7.14 A tang and clevis joint as used on the SRB.

out and the shuttle climbing past—so the next emission was always from the point where the joint was, the booster having climbed past the smoke emitted from the previous puff.

As the SRBs are constructed of several segments it is necessary to join these together in a manner that prevents the hot exhaust gases from escaping. Some of the joins are resealed at the factory when new solid propellant is packed into the shell. Others, the so-called 'field joints', are resealed when the boosters are delivered to NASA. The field joints consist of a tang (male part) at the bottom of the higher segment and a clevis (female part) at the top of the lower segment (see figure 7.14). The small remaining gap between the joints is sealed by a pair of rubber O rings running round the circumference of the booster. These rings are approximately 8.2 m in circumference and 6 mm thick. The physical properties of the rubber used in the manufacture of these rings are very important, as they have to be able to expand with the joint to maintain the seal under load.

Analysis carried out by the engineers of the Morton Thiokol Company responsible for the joint seals showed that during launch the pressure of the exhaust gases would cause the casing of the SRB to bulge outwards between the joints. This effect did not happen in the region of the joints as the casing was naturally thicker there. However, it did have the effect of opening up a gap between the tang and clevis of the joint (an effect known as 'joint rotation'). The engineers had observed on previous flights that the O rings were occasionally blackened and burnt by gases leaking through this gap. On the morning of the flight several Thiokol engineers recommended that the launch be postponed. They were worried that in the intense cold the rubber of the O rings would stiffen and not expand fast enough to seal the gap when joint rotation happened.

- About 37 seconds after launch the Challenger started to feel the effect of high-altitude winds pushing it off course. This continued until about 64 seconds into the flight. The on-board systems compensated for the winds and the SRBs altered thrust direction as instructed. Conditions were such that the system was more active than on any previous flight.
- As the shuttle accelerated to the speed at which the atmospheric drag on the ship would peak, the main engines and SRBs reduced thrust, as designed, to smooth the transition into the cleaner air flow beyond this speed. After this the engines were throttled up to 104% of nominal thrust. At this time the first signs of a small flame were later seen on film. It can first be detected at 58.788 seconds into the flight using computer enhancement of the image. The flame is in the vicinity of the same joint from which the smoke was seen earlier in the flight. Presumably whatever had resealed the joint earlier had now failed.
- On the very next frame the flame can be seen without enhancing the image. By 59.262 seconds it had established itself as a continuous plume. Information relayed to ground by on-board sensors at 60 seconds into the flight showed a difference in pressure between that inside the right-hand SRB and that inside the left, indicating that a leak was reducing the pressure.
- As the plume grew bigger the airflow over the shuttle deflected it so that it was playing on one of the lower struts that attach the SRB to the external tank and onto the external tank casing itself (figure 7.15).
- At 64.660 seconds the flame changed shape and colour, indicating that it had burnt a hole in the external tank and that it was mixing with hydrogen that was leaking out. Telemetry also recorded a drop in helium tank pressure, confirming that a leak had started. Some 45 milliseconds after this a bright glow can be seen reflected in the black tiles on the underside of the orbiter.
- Round about 72.20 seconds into the flight the lower SRB support structure failed and the right-hand SRB rotated downwards about the upper strut. The impact dented and burnt the wing of Challenger.
- At 73.124 seconds a pattern of white vapour developed around the edge of the lower dome of the external tank. Shortly after this the whole aft dome fell away, releasing massive amounts of liquid hydrogen. This created a sudden forward thrust of about 12.3 million newtons, forcing the remains of the hydrogen tank upward towards the LOX tank above. At the same moment the rotating SRB struck the inter-tank structure and the lower part of the LOX tank. Both of these structures failed at 73.137 seconds into

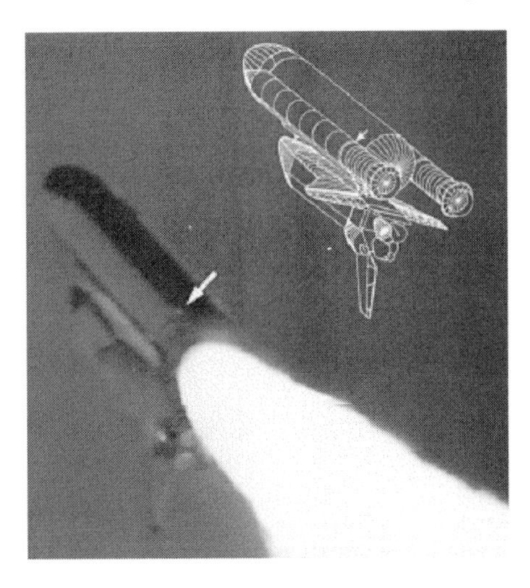

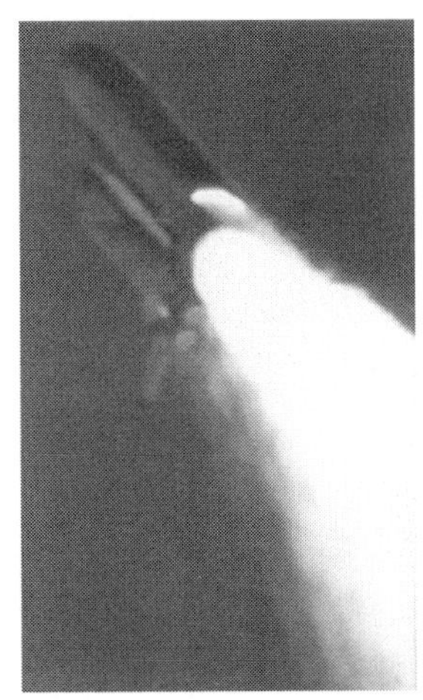

Figure 7.15 Exhaust from the failing SRB joint starts to play on the external tank.

the flight—confirmed by the white vapour seen emerging from this region at that time. Within a few milliseconds of this the liquid oxygen and hydrogen leaking from the tanks started to burn with a rapidity that was almost explosive.

- Challenger had reached the moment in its trajectory when it was travelling at Mach 1.92 at an altitude of 14 km. It was enveloped in the burning cloud of propellant and the orbiter broke apart under the stress. The hypergolic propellants used by the RCS systems burst free and the reddish-brown colours they produced on burning can be seen at the edges of the main fireball. The film also shows several large pieces of the orbiter falling out of the fireball. A wing, the tail section with the engines still burning, and the forward fuselage (trailing a mass of pipes and wires pulled from the payload bay) can all be identified on the film.

The investigation concluded that the accident had been caused by a failure of one of the field joints on the right-hand SRB. The problem of joints not sealing properly under joint rotation had been exaggerated that morning by the very cold conditions, which had stiffened the rubber in the O rings to the extent at which they could not expand fast enough to maintain the seal.

In an appendix to the report Feynman stressed the importance of listening to the engineers responsible for maintaining the system who had urged caution with regard to several areas of shuttle operations. He was especially scathing of the system used for estimating the likelihood of component failure. The official estimations had placed the chance of a severe problem occurring as 1 in 100 000 launches—meaning that a shuttle could be launched every day for 300

Figure 7.16 The final moments of Challenger.

Table 7.1 Shuttle designations.

Name	Date first flew	
Enterprise	1977	Used in landing tests, now in the Smithsonian Museum at Dulles Airport outside Washington
Columbia	April 1981	Flew first 5 missions, now modified to support long-duration missions (up to 16 days)
Challenger	April 1983	Built as a vibration test vehicle, then upgraded to become the second shuttle to fly in space. Challenger exploded with the loss of its crew on 28 January 1986
Discovery	August 1984	
Atlantis	October 1985	
Endeavour	May 1992	Built to replace Challenger

years without a major problem occurring. More junior engineers were estimating the chances as being more like one in a few hundred. However, the message was not being heard in the hierarchy of NASA.

7.6 What follows the shuttle?

Now that assembly of the international space station has started, the space shuttle will finally come into its own as a flexible, re-usable launch platform capable of acting as a base for engineering applications. Over the next few years the shuttle will be used to carry a large number of components into orbit. Indeed, most of the shuttle flights between now and 2004 will be related to the ISS. However, the orbiter fleet has a finite life expectancy and NASA is already researching a radical new approach to the concept of a re-usable launch vehicle (RLV).

For a craft to be fully re-usable, every component has to be capable of re-entry and recovery. The best way of achieving that is to integrate every part into a single system. The single stage to orbit (SSTO) concept is to use an all-in-one craft that will launch like a rocket and land like an aeroplane. The idea is to take an orbiter-like craft and give it the power to lift directly into orbit without an external tank or extra boosters. In order to manage this a much greater thrust must be achieved (remember the reasons for staging outlined in chapter 3). This can be done from a combination of more efficient engines and reduced mass in the launch vehicle.

In the early 1990s NASA started a programme with the ambitious aim of reducing the cost of carrying payloads into orbit from $10 000 per pound (mass) to $1000 per pound within 10 years. The programme is designed to encourage industry to take up the challenge in cooperation with NASA centres on a cost-sharing basis.

The first step in the programme was taken by the McDonnell Douglas Corporation with the design and flight testing of the DC-XA (*Clipper Graham*). This was a technological demonstrator that proved one of the most important technological advances needed for a real SSTO—reducing the mass of the launch vehicle by using lower-mass materials in the propellant tanks. *Clipper Graham* used a composite LH_2 tank, an aluminium–lithium LOX tank, and composite fuel lines. Before it was destroyed in an accident at the end of a test flight, the craft repeatedly demonstrated the ability to take off vertically, move to one side, hover over the landing site and, by throttling its engines, descend again. Currently this work is extending in the X-33 and X-34 programmes.

X-33

In 1996 NASA decided to work in cooperation with Lockheed-Martin and is providing funds of approximately $1 billion over 42 months to build and fly the X-33 SSTO prototype. This cooperative agreement is a mechanism by which

Figure 7.17 A graphic showing the design of the X-33.

the US Government can share with industry the risks and benefits of working towards a common goal—in this case low-cost access to space. In such a scheme no profit is made by the industry. Lockheed-Martin is contributing $200 million to the X-33 programme.

The selected team consists of Lockheed-Martin, Rocketdyne (engines), Rohr (thermal protection systems), Allied Signal (subsystems), and Sverdrup (ground support equipment), and various NASA and defence department laboratories.

X-33 is a sub-orbital prototype about half the size of a genuine SSTO craft that could carry a payload. It has been designed to demonstrate some of the technology that might be used in the full-scale version (named *VentureStar* by Lockheed-Martin and due for development after the turn of the century).

The X-33 will take off vertically and glide to a landing. The systems will be fully automated (requiring no pilots on board), and although the craft will not reach orbit, it will be capable of following a flight path that will test all the critical parts of an SSTO in realistic conditions.

One of the more radical pieces of technology used in the X-33 is the Linear Aerospike engine developed by Rocketdyne. This engine does not use a conventional nozzle at all. Instead the exhaust cascades either side of a curving sheet of metal (like water over the edge of a waterfall) called the nozzle wall.

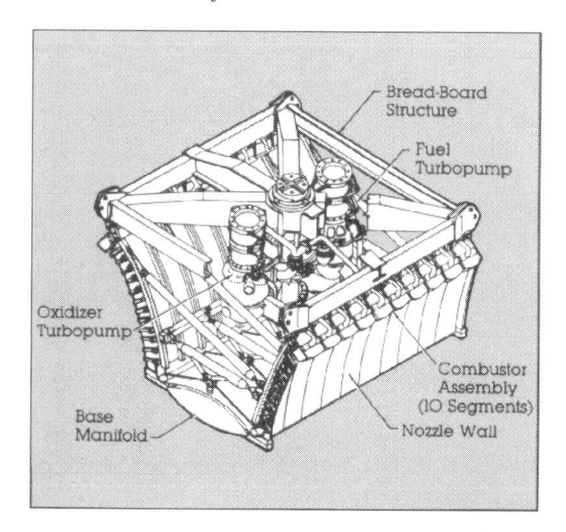

Figure 7.18 Top, the Linear Aerospike engine. Lower left, a Linear Aerospike engine on a test frame; lower right, test firing the Linear Aerospike.

On each side the exhaust is constrained by the nozzle wall on its inner side and open to the atmosphere on the outer side.

This design confers two great advantages.

Firstly, with the exhaust flow open to the atmosphere it is able to freely adapt to the atmospheric pressure in a manner that no engine bell can. This allows the engine to perform equally well at a variety of different altitudes as the exhaust flow pattern is self-compensating. The design works best if the two nozzle walls continue down and join together, forming a spike under the engine (hence the name). However, the Rocketdyne engineers discovered that they could simulate this quite accurately by allowing the exhaust gases from the turbopumps in the engine to exit at the base of the truncated spike (the base manifold). This flow of gas keeps the main engine exhaust separated in just the way that continuing the nozzle walls into a spike would, with much less weight and structural problems.

Given that opening the exhaust flow to the external pressure produces a performance advantage, we naturally wonder why the nozzle wall is needed at all—why not just open the whole flow to the external environment? The reason is that in this engine the thrust is produced by the pressure of the exiting exhaust gas acting against the nozzle wall. If there was no nozzle wall, there would be no thrust produced! (Note that there is no combustion chamber in quite the same way as there is for a conventional engine.)

The second advantage that the aerospike has over a conventional design is the ability to remove the need for the engine to gimbal. The aerospike engine sits at the back of the X-33 and runs along the width of the craft. The same steering effect can be produced by regulating the exhaust flow over the upper or lower nozzle wall. A greater flow over the top produces more thrust there and tends to tip the craft downward, and vice versa.

Once X-33 proves itself on its first test flights, the programme will move to demonstrating a 99% reliability of the craft with a turn-around time that can be measured in days, not weeks as with the space shuttle.

There is considerable interest in X-33 and the NASA SSTO programme in general. For some while there has been a camera in the X-33 hangar periodically taking pictures of the assembly for broadcast on the Internet. No doubt the same sort of interest will be shown in the test programme.

At the time of writing, the aerospike engine and the composite hydrogen tanks are undergoing final testing before installation on the X-33, which should be rolled out for its first test flight in January 2000.

X-33 specifications

Length: 21 m
Width: 23.5 m
Take-off weight: 129 tonnes
Propellant: LH_2/LOX
Propellant weight: 95 tonnes
Main propulsion: (2) j-2s Linear Aerospikes
Take-off thrust: 1 822 535 N
Maximum speed: Mach 13+

Figure 7.19 A graphic showing the design of the X-34.

X-34

On 28 August 1996, NASA awarded a contract to the Orbital Sciences Corporation (OSC) for the design, development and testing of the X-34. This is intended as a technology test bed—a vehicle that can be used to demonstrate the viability of some of the technology required for an SSTO vehicle. It will act as a link between the *Clipper Graham* and the X-33.

Unlike the X-33, the X-34 will be dropped from a modified aeroplane. Its engines will then start to carry the craft to the planned altitude and speed, after which it will coast, re-enter and land automatically. The key technologies incorporated into X-34 include the following.

- *Composite airframe structures, composite re-usable propellant tanks, cryogenic insulation, and propulsion system elements.*
- *Advanced thermal protection systems and materials.*
  New ceramic thermal protection systems which are far less expensive than the current shuttle ceramics.
- *Low-cost avionics including differential GPS and integrated GPS/INS.*
  Low-cost flight-proven avionics including differential GPS (global positioning systems). The final RLV vehicle will depend heavily on automatic navigation, guidance and control for autonomous landing, so it is important to demonstrate that performance, simple checkout and testing, and operational cost objectives can be met.
- *Integrated vehicle health monitoring system.*
  An integrated vehicle health monitoring system which, together with rapid software re-programming, will it make it possible to re-use the vehicle in a short timescale. This is a major objective for a new RLV as it will cut the cost of lifting payloads dramatically.

7.7 Other ideas

According to a study carried out by the accountancy firm KPMG Peat Marwick in 1996, world-wide commercial interests spent more money in space that year than governments did. It seems likely that this trend will continue over the next few years. At least one estimate suggests that 1200 telecommunications satellites alone will be launched between 1998 and 2007. With this explosion in private-sector demand for launches into space, the pressure to bring down the cost of lifting payloads into Earth orbit will increase. This demand has led to some extraordinary ideas:

- Boeing's commercial space division has teamed up with RSC-Energia in Moscow and Kvaerner Maritime from Oslo to modify an oil rig for use as a launch platform that can be towed to sites that suit the orbit that is being aimed for.
- NASA and the USAF have sponsored a series of experiments in which a 10 kW IR pulsed laser has been used to propel a small model into the air. The 15 cm diameter lightcraft is cone shaped with the open end facing downwards. Reflective surfaces on the inside focus the light from the laser into a ring that heats the air under the craft to five times the temperature of the surface of the Sun. The air then expands explosively, providing thrust. In a five year programme the researchers aim to accelerate a 1 kg microsatellite into orbit using a 1 MW laser and a few hundred dollars of electricity.
- The USAF has a four year $48 million research programme into using solar power for the final stage of a rocket. The idea is to use a mirror to direct the sun's light onto a graphite block containing liquid hydrogen. The sunlight heats the block to 2100 °C vaporizing the liquid which then expands as a gas to provide the thrust. Such a system could boost a satellite from low Earth orbit into geostationary orbit by thrusting for 3–8 weeks. It would have to be lifted into low orbit by conventional means, but the estimates are that this system could still save tens of millions of dollars on each launch.

These are just a few of the ideas that are being explored for reducing the cost of exploiting space. Not only should such developments pay dividends in the commercial sector, they should also help in the exploration of Mars and the rest of the solar system.

Notes

[1] Various people have suggested that rather than being left to burn up these tanks could be left in orbit. When enough of them were in place they could be marshalled together to construct an orbital hotel.

[2] On the STS-71 mission to Mir in July 1995, seven people flew into orbit, and eight came back. That is the largest number of people that have ever flown in the shuttle.

[3] There is a hierarchy of computer languages. Computer chips understand only very simple instructions and software written in their language is very complex and hard to follow. Computer languages that are of a higher level (such as Basic, FORTRAN, C, Visual C++, etc) are more easily understood by humans, but have to be translated into machine language before the chips can run them.

[4] Curious choice of name for a language used to programme a spacecraft's computers, isn't it?

[5] Sometimes the launch is held even though the weather at the Cape is apparently perfect. This can be because the weather at one of the emergency landing sites is not suitable.

[6] NASA has visas on hand ready to ship them rapidly to the appropriate country should the crew be forced to make an emergency landing.

[7] These are the figures for a typical mission. Clearly some have slightly different first orbits due to the differing requirements of the mission.

Intermission 5

The politics of Apollo

On 25 May 1961, John F Kennedy made a speech to the American Congress committing his government to landing a man on the Moon before the end of the 1960s. It is an inspiring speech full of passion and promising that this 'greatest of all adventures' would rebuild the country's confidence, which had been badly dented by Soviet successes in space.

> 'No single space project in this period will be more exciting, or more impressive to mankind, or more important for the long-range exploration of space; and none will be so difficult or expensive to accomplish . . . '

By the time Kennedy delivered his congressional speech on 'Urgent National Needs', only one American had flown in space. Alan B Shepard's 15 minute mission grazed space on 5 May 1961. This was about a month after a Soviet mission in which Yuri Gagarin made one complete orbit of the Earth (12 April 1961). Although Shepard's flight took place only a month after Gagarin's, and if it had not been for the understandable caution of NASA Shepard could have flown first[1], the achievement was technically short of what the Soviets had accomplished. The first American to orbit the Earth was John H Glenn, who carried out three orbits on 20 February 1962.

The first satellite to be placed in orbit had also been Soviet. Sputnik 1 passed over the United States in 1957, emitting a beeping radio signal and sounding eerily like the countdown to a nuclear bomb.

Given the state of the American space programme at the time when Kennedy made his promise, it is hardly surprising that some people at NASA were privately horrified by the timescale imposed of 'before this decade is out'.

In taking over the presidency from Eisenhower, Kennedy inherited a manned programme (Mercury) that was struggling to compete with the Soviet achievements and a scientific community that was deeply divided over the value of manned space exploration. Both Eisenhower's and Kennedy's chief scientific advisors felt that the results of manned flight could not compare with the likely benefits of an unmanned satellite programme. Indeed Eisenhower himself, in his departing budget speech (18 January 1961), said that more work would be needed to establish if there was any benefit in extending the manned space effort beyond the Mercury flights.

NASA engineers were confident that they could land a man on the Moon, given sufficient development time. Indeed they were convinced that a lunar landing was the only sensible conclusion to the current phase of manned space exploration. Much scientific research could be done with unmanned probes, and satellite technology was set to revolutionize communications and weather forecasting, but for exploring space, NASA was convinced that people had to fly. In October 1960 they issued contracts to three aerospace firms to study the feasibility of a lunar mission. However, NASA's research and planning would be in vain without substantial funds being made available and that required Presidential and Congressional backing. Trying to justify such an expenditure in terms of pure science was not going to work. Human exploration of the Moon, no matter how poetic and important to the human spirit, was not going to reap scientific rewards that could not be obtained by automatic probes. After all, the people being trained to fly were not scientists. In the end NASA's ambitions had to be realized on non-scientific grounds. The climate of competition with the Soviets was just what was needed to put pressure on the President.

Whatever the scientific rights and wrongs of the argument, those in favour of unmanned missions did not take into account the role of public and political opinion. On the one hand the public could not get as excited as the scientists about lobbing unmanned satellites into orbit. On the other, politicians and media moguls could see highly successful Soviet space missions and very public and embarrassing American failures. They were pressing for the Americans to beat the Soviets as a matter of pride. This pressure mounted to a head when Gagarin made his historic flight.

Kennedy, as President-elect, called on his Vice President-elect, Lyndon B Johnson (who was an enthusiastic supporter of the space programme) to conduct an investigation into the nation's space efforts, find a project that would produce 'dramatic results' and demonstrate the USA's superiority in space. Johnson's report was presented on 8 May 1961. It recommended uprating the space programme, commissioning a new rocket to lift heavier objects into space and committing NASA to landing a man on the Moon. The result was Kennedy's speech to Congress. Space was not the only issue covered, but it will always be remembered as the moment that America decided to go to the Moon.

Congressional support was not assured, but they turned out to be solidly behind the idea. Kennedy asked to increase Eisenhower's $1.1 billion budget for the space programme by $675 million. He got virtually all of that money and the Apollo programme was underway.

With the assassination of Kennedy in 1963, Johnson took office and subsequently won the 1964 election. Johnson was in power during the build-up of the Apollo programme and was a staunch supporter of NASA's budget requests. Indeed it has been commented that Johnson was the only American president to be genuinely interested in the space programme, rather than tolerating it for political expediency. By 1968 Johnson's involvement in the Vietnam War had turned public opinion against him and he did not contest another term of office. Nixon won the 1968 election. In July of the following year Apollo 11 landed on the Moon and Nixon congratulated the astronauts personally over a radio link to the White House while they stood on the surface.

Almost immediately after the Apollo 11 astronauts landed back on Earth, public interest in the space programme began to decline. There was a brief re-surfacing during the nerve-wrackingly hazardous Apollo 13 mission, but the tide had turned. After a period of guaranteed presidential support NASA was now working in a different climate. Furthermore, Apollo was not the only programme that NASA wished to develop. The Mariner series of probes visited Mars and Venus, sending spectacular pictures emphasizing that manned exploration was not the only way of bringing science back from space. Planning for the Viking mission to soft-land a probe on Mars, the Skylab Space Station and the early development of the space shuttle were also taking place during the Apollo years and NASA had to provide funds for these as well. In 1970 the NASA operational budget was of the order of $3.8 billion, but now Congress was in the mood to make cuts.

The problem NASA faced was the perception that Apollo had been staged primarily as a race against the Soviets. The scientific benefits of the programme were dubious and had not been emphasized. The race had been won. In order to justify the continual staging of flights, the science now had to be stressed.

In 1970 NASA took to congress a request for a $3.33 billion budget for 1971, a $500 million cut on the previous year, and a suggestion to cancel one of the Apollo missions—Apollo 20. In the end they had to settle for something more like $3.27 billion. More hard decisions had to be made and on 2 September 1970 NASA announced that two more Apollo flights were to go. This triggered an uproar in the press. Interestingly enough, some of the voices that had originally been outspoken against staging Apollo were now arguing that it should not be cut back. The *New York Times* pointed out that the savings made by cancelling three Apollo flights amounted to no more than 2% of the NASA budget and 0.25%

of the total money invested in the Apollo programme thus far: 'this decision can only vindicate the critics who have insisted that Apollo was motivated by purely prestige considerations, not scientific goals'.

Each Apollo flight represented a cost of $20 million, compared with the $25 billion that had already been invested. All the spacecraft and rocket boosters had already been built (in the end, one of the rockets was used to launch Skylab and the other two are on display at the Kennedy and Johnson Space Flight Centers). However, NASA wanted to support a post-Apollo programme and there was no valid alternative to the cuts. Interestingly, in mid-September 1970 a letter was sent from 39 Apollo scientists to the Chairman of the House Committee on Science and Astronautics. The point was made in the letter that the money being saved from the Apollo programme, an approved and evidently successful sequence of flights, was being used to foster 'an as yet unapproved program for whose scientific value there is no consensus and whose purpose is unclear'—the space shuttle. The reply sent by the Chairman also makes strong points: 'Had your views on the Apollo program been as forcefully expressed to NASA and the Congress a year or more ago, this situation might have been prevented'.

The last three Apollo missions were dead. Since that time, no pressing scientific need to return to the Moon has emerged. If Apollo had been a scientific mission, then it probably would never have flown. These days the priorities are different and it is hard to argue against that.

There is another very telling comment in the film Apollo 13. Lovell and his wife are sitting in their garden gazing at the Moon. It is the night of Armstrong and Aldrin's first walk on the Moon. Lovell is wistfully observing the Sea of Tranquillity: 'from now on we live in a world where man has walked on the Moon—and it's not a miracle. *We just decided to go*.' The context for that decision was exactly right in the 1960s. That context has now gone.

Notes

[1] NASA flew a test mission of the Redstone rocket that would carry Shepard on 31 January 1961. The test flight carried a chimp called Ham and although it returned Ham alive to the ground, several technical problems arose. As a result a second test flight occurred on 24 March, which was completed perfectly. Shepard had campaigned to fly the second mission confident that all the technical problems had been fixed. As he commented later (referring to the competition with the Soviets) 'we had 'em by the short hairs, and we gave it away'.

Chapter 8

Mars

In this chapter we will explore an interesting new plan for a manned mission to Mars. Mars Direct, as its principal architect, Robert Zubrin, has called it, represents an intelligent and cost-effective method for opening up Mars to sustained exploration and future colonization. In the heady days of optimism surrounding the Apollo programme, people started talking about Mars missions as if they could be accomplished within the decade after Apollo. Yet, thirty years later there is still no definitive commitment to a mission.

8.1 The modern scene

On 20 July 1989 President George Bush, in an echo of Kennedy's speech, committed America to landing a man on Mars. Twenty years to the day after the Apollo 11 landing, and flanked by the three astronauts who took part in the mission, President Bush launched a ten-year initiative that would build up to a Mars mission via a space station and a return to the Moon. Thus was born the Space Exploration Initiative (SEI). News of the initiative carried to England, but then an ominous silence seemed to descend over the whole project.

NASA convened a committee to carry out a feasibility study for SEI. The committee returned a complex strategy[1]. They suggested modifying the planned *Freedom* space station[2] and constructing a lunar base. Together these facilities could build a 1000 tonne spacecraft to travel to Mars powered by newly developed engine technology (see chapter 3). The assembly could not be carried out on Earth, as the completed craft would be far too heavy to lift into orbit.

The estimated cost for all this ran to $450 *billion*. No wonder the whole thing went rather quiet. The political context for allocating such a huge amount of funding simply did not exist at the time, and although there is evidence that the tide may well be turning, such an expenditure simply cannot be contemplated even today.

NASA and the whole space programme had a public mauling in the late 1980s and early 1990s due to a series of highly embarrassing failures—such as the Challenger disaster and the wrong shape of the Hubble Space Telescope's mirror. The space shuttle, originally promised as being the key to accessing Earth orbit, was far less cost effective than NASA wanted to admit. The whole organization surrounding space exploration seemed to be beaurocratic, blinkered and wasteful of taxpayers' money.

Interestingly, the recent tide has been turning back. Carrying out the highly ambitious repair to the Hubble telescope, in the full glare of public scrutiny, was a marvellous success. The availability on the Internet of the glorious photographs that the telescope now produces started a trend that has been effectively exploited in recent years. Public interest and participation in the Galileo mission to Jupiter and its moons, and the Mars Pathfinder and Global Surveyor missions, exceeded all expectations. Once again, people could have access to information almost as fast as the scientists received it themselves.

NASA was also handed a golden opportunity to generate interest with the Shoemaker–Levy comet impact on Jupiter. The scientists and public relations people involved have handled all of these examples with great skill. NASA has learned many lessons since Challenger. It is building interest and support for a Mars mission slowly, carefully and effectively. The possible discovery of evidence for fossilized microscopic life in a sample of Martian rock has increased the support for a manned mission to the red planet.

It is possible that by the turn of the century public interest in a manned Mars mission will have reached a level that will enable it to be carried out. However, the plan will have to be a great deal cleverer (and so cheaper!) than that produced by the Bush committee. Fortunately a growing band of enthusiasts, both amateur and professional, have been forming an underground network dedicated to developing mission plans. As always when such things are brainstormed by enthusiasts free from political pressure or ties to particular companies and ideas, the results are radical, practical and inventive. The Mars Direct plan encapsulates all these virtues.

8.2 Mars Direct

In devising the Mars Direct plan, Robert Zubrin's starting point was the assumption that there is no point in going to Mars simply to plant a flag. Any mission to Mars had to be worthwhile and that involved two things—searching for life, or the evidence for past life, and exploring Mars with a view to future colonization. Both of these aims required astronauts to remain on the surface for substantial periods of time. The problem with the traditional plan was that the proportion of time spent on the surface compared with the time taken to travel to Mars was far too small—for the simple reason that the spacecraft had to carry all its supplies with it.

The traditional mission plan reads very similarly to that of Apollo. Scale is the only real difference, but that is a crucial one. Given that the vehicle used would have to support a team of astronauts for a mission lasting about two years, and given the requirement to carry all the supplies, fuel and oxygen needed for a return journey, the design called for an enormous 'Battle Star Galactica' (as Zubrin calls it—revealing poor taste in science fiction) spacecraft. This in turn forces upon the mission designers the need to assemble the craft in orbit. Not only is it doubtful that such an enormous engineering exercise can be carried out in the near future, the plan also forces upon the designers the need for a fully functioning manned space station to act as a staging point and shipyard. Such issues rapidly escalate the cost towards the $450 billion mark.

While there are many good reasons for the construction of a permanent space station in Earth orbit, it is dangerous to mix two ambitious engineering enterprises together. One of the dangers is that supporters of the space station project see the Mars mission as a way of generating support and money for their pet project. At this point the issue becomes self-fulfilling, and it is difficult to separate what is necessary and desirable to achieve one end from the ambitions and interests of those involved with achieving an independent goal.

The constraints imposed by having to carry the mission supplies along obviously result in a need to minimize the total mission time. For this reason many NASA plans have worked on the basis of one of the journey's legs using either a Venus gravity boost (as discussed in chapter 4) or a large amount of propellant to increase the ΔV.

As Zubrin has pointed out, this sort of plan minimizes the total mission time largely by reducing the time spent on the surface and, as in the case of the Venus fryby, increasing the crew's dose of cosmic rays. If a means could be found to support the crew while they are on the surface, then a more conventional Hohmann transfer type of mission becomes possible. This would enforce a longer stay on the surface (to wait for a return window to open), which is actually desirable. Not only would this allow more science to be done on the surface (crucially the search for evidence of life), but the atmosphere of Mars provides some shielding from cosmic rays. In Zubrin's view the correct approach is to minimize the time in space and maximize the time on the surface.

In 1990 Zubrin and David Baker worked out the substantial features of the Mars Direct plan. The breakthrough came from a modification of an idea first put forward by Jim French in an article for the journal of the British Interplanetary Society. French suggested landing an automated propellant manufacturing plant on Mars before the crew arrived. The plant could then use the local Martian resources to produce a stock of propellant for the crew to use on the return journey. As the manned mission would not need to take the return propellant with it, the payload could be reduced and more supplies carried along. The problem was landing the manned mission 'within a hose length' of the propellant plant. That was a degree of flying that not even the expert pilots at NASA would wish to contemplate—especially as they would not be coming home if they did not succeed!

The Mars Direct modification to the idea is very simple but highly effective. You do not land a propellant manufacturing plant ahead of the crew—you land a whole Earth return vehicle equipped with such a plant. While on the surface the vehicle uses its in-built propellant plant to refuel itself for the return journey. When this has been done it signals its readiness to Earth. The manned mission then departs, but in a much smaller spacecraft than would otherwise be needed as the crew is not using it to return to Earth.

There is no need for a Mars lander to descend to the surface from a mother ship that remains in orbit. Landing had always been a problematical feature of mission plans anyway. Not only does carrying along a Mars lander increase the mission's weight, but it opens up the question of what to do with the mother ship. Either a crew member is left on board to be exposed to cosmic rays and prolonged zero g, or the craft is left empty and the crew take the risk of it not being in perfect condition when they return. These factors always argued for a short stay on the planet (as in the Apollo missions).

The alternative to taking a lander along is to use the main vehicle to land on the surface (as in the direct-flight Moon plans). This raises the same problems as it did for Apollo—can the crew risk a potentially rough landing in the same ship that they will be using to return to Earth? The advantage of the Mars Direct plan is that the crew knows *before they leave home* that there is a fully functioning return vehicle that has survived landing on Mars waiting for them.

Another advantage of such a mission profile is that the two craft making separate journeys to Mars are both much lighter than a combined return vehicle would be, making it easier to aerobrake them into orbit or to a landing.

In summary, then, Mars Direct works like this. The first launch is to send an unmanned ship directly to Mars from Earth's surface (no assembly in orbit). This craft, with a mass of about 45 tonnes, takes a Hohmann-type transfer orbit and arrives on Mars to aerobrake to the surface. It carries supplies, equipment and a chemical plant. Once on the surface the chemical plant on board uses Mars' atmosphere to manufacture propellant for its return journey. Having done this the craft signals back to Earth that it is ready to receive a crew. Using robots similar to the already proven Mars Pathfinder, the team on Earth explore the vicinity of the return vehicle, choosing a safe place to land. Having selected one, they use the robot to deposit a radio beacon to guide them to the surface.

Once ready for the next stage of the mission, another launch takes place to send a second return vehicle, identical to the one already on Mars, on its 250 day Hohmann transfer journey. A few weeks later a third launch boosts a manned version of the craft on a higher ΔV transfer orbit so that the crew take about 180 days to arrive at Mars. This is more easily done in the lower-mass craft that they will be using. Their ship is a habitation module (Hab) 5 m tall, 8 m diameter cylinder with two decks and enough supplies on board for an 800-day mission.

On the way they disconnect their ship from the third stage of the booster rocket, but rather than ejecting the spent stage into space they leave it attached to their craft by a long tether. The stage then forms a counter weight, allowing them to spin the combination about a combined centre of gravity, giving some artificial gravity along the trip. Once at Mars they cut the stage away and aerobrake to a landing somewhere within walking distance of the return vehicle (remember that Apollo 12 landed within 160 m of the Surveyor probe and that did not have a homing beacon).

One of the significant problems faced by mission designers is what to do if a problem should develop on the spacecraft. The Apollo 13 mission turned out well as the crew could survive in the lunar module and the distance they had to travel to return to Earth was much less than that facing a Mars expedition team. With a mission following the Mars Direct plan the crew now have several options open to them should a problem arise.

- Once they are nearer to Mars than the Earth their best bet is to continue on to Mars, where there is an inhabitable craft waiting for them on the surface, rather than trying to return to Earth.
- If they are unable to land within walking distance of the return vehicle they can use the rover that they carry with them to drive to its location.
- If they are further away than a rover drive they can guide the second return vehicle, which is still on the way, to land near to them.

With ample supplies on their craft and the return vehicle they are able to remain on the surface for 500 days, waiting for a low energy return window to open (doing some science while they are there). In the meantime the second return vehicle has arrived at some other spot and is now preparing propellant for the second manned mission that can start out as soon as it is ready.

Mars Direct offers the opportunity for a sustained and systematic exploration of the surface by a sequence of such missions. And it could all be done with a booster rocket no different from the Saturn V used for Apollo.

8.3 Chemistry on Mars

Clearly the whole of the Mars Direct plan hinges on the manufacture of propellant on the surface of Mars, a task that sounds more ambitious than it actually is. Mars' atmosphere is very thin, a little less than 1% of the pressure at Earth's surface, and composed almost entirely of carbon dioxide (CO_2).

The plan is to ship the Earth return vehicle to Mars carrying six tonnes of liquid hydrogen. It is relatively easy to keep this cool enough while travelling through space and about 15% extra can be allowed on launch for 'boil off' *en route*.

Once on the surface the liquid hydrogen (stored in insulated tanks—the technology for doing this already exists) is reacted with carbon dioxide from the atmosphere:

$$CO_2 + 4H_2 \rightarrow CH_4 + 2H_2O$$

producing methane (CH_4) and water (H_2O). The methane is stored in liquid form (perhaps by cooling it in contact with the liquid hydrogen). The water is also stored, but it is then decomposed electrically:

$$2H_2O \rightarrow 2H_2 + O_2.$$

The resulting hydrogen is used to react with more carbon dioxide and the oxygen is liquefied and stored. Clearly the oxygen is the oxidizer needed for the return journey, and the methane (lighter fluid) is the fuel. Burning methane in oxygen with a mixture ratio of 2:1 produces a specific impulse of 340 seconds. It would be better to use a ratio of 3.5:1, which would require more oxygen than could be produced in the reactions detailed above. This mixture would give a specific impulse of more like 380 seconds. A variety of different possibilities are available for making the extra oxygen, such as the direct decomposition of carbon dioxide. Whatever process is chosen, more than enough oxygen can be produced to provide oxidizer for the fuel—and to store for the crew's use (as well as water) when they arrive.

In 1993 Zubrin was part of a team that designed and built a working plant that could produce 400 kg of propellant (for a robot probe designed to return samples of Martian soil to Earth) with a mass of 20 kg and requiring 300 W of electrical power. The system functioned with 96% efficiency.

There is little doubt that such a system could be scaled up for use in Mars Direct.

8.4 The NASA reaction

In 1992 NASA broke with its traditional paradigm and adopted a variant on the Mars Direct plan as their baseline method for getting a manned mission to Mars. Mars Semi-Direct, a compromise proposal worked out by Zubrin and Dave Weaver of the Johnson Space Center, works on three launches per mission. One sends an unmanned return vehicle to the surface of Mars, where it manufactures propellant. A second unmanned launch sends a fuelled Earth return vehicle into orbit about Mars. The third launch sends the crew. They land on the surface of Mars near to the craft, which has made enough propellant for a return to Mars orbit. When they wish to return the crew use this vehicle to dock with the Earth return craft, which had brought its propellant from Earth, in Mars orbit. They transfer to the return craft and set off for home.

In 1993 the plan was handed to a team of representatives from all the NASA facilities. They fine tuned the idea and then passed it on to the same group that had costed the 90-day report. The answer came back at a much more realistic $50 billion.

What are the objections to a complete Mars Direct plan? Primarily that the mission hinges on the manufacture of a large amount of propellant on the surface. Mars Semi-Direct uses a more familiar mission profile and a smaller amount of propellant is needed on the surface.

However, the concepts of using propellant manufactured on Mars and sending smaller craft, in sequence, that do not require assembly in Earth orbit are now part of standard NASA thinking. It remains to be seen how attitudes evolve over the next decade.

8.5 Mars in the near future

NASA has instigated an ambitious programme of unmanned probes to Mars every two years (when a launch window opens) for the next seven years.

Figure 8.1 Artist's impressions of various events during a manned Mars mission using the Mars Semi-Direct mission profile. Top left, landing a hab module on the surface of Mars; top right, crew working near to the Mars ascent vehicle; lower left, lift off from Mars in the Mars ascent vehicle; lower right, rendezvous in Mars orbit with the Earth return vehicle.

Mars Climate Orbiter

This probe was launched successfully on 11 December 1998. It is currently *en route* to Mars and is due to arrive on 23 September 1999. On arrival it will carry out a burn with a ΔV of 1.2 km s^{-1} which will place it into an elliptical polar orbit. It will then use its solar panel as an aerobrake until a lower elliptical orbit is achieved by about 22 November 1999. Once the Mars Polar Lander has arrived on 3 December 1999, the Orbiter will act as a relay station transmitting information gained from the Lander as it passes over the landing site for 5 or 6 minutes ten times every Martian day (slightly longer than an Earth day). On 3 March 2000 it will switch roles and start mapping the surface and observing weather patterns for one Martian year until 15 January 2002. At this time it will then switch into its final role, move to a stable orbit and act as a relay station for the Mars 2001 missions which should be arriving.

DS2

This is the second of the New Millennium probes first mentioned in chapter 3. It was launched along with the Mars Polar Lander on 3 January 1999. On 3 December 1999 the Polar Lander will separate out from the cruise stage of the probe, and shortly after the two DS2 probes will detach as well. The DS2 probes have no propulsion systems of their own, but they are designed to properly orient themselves as they pass through the Martian atmosphere. They will smash into the surface in the South Polar region with a speed of between 160 and 200 m s^{-1}. On impact their aeroshells (designed to protect them during re-entry) will shatter and the probes will split in two. Part of each probe will remain on the surface while another section penetrates 0.3–1.0 m under the surface (depending on the nature of the material), relaying information to Earth via the surface. The surface parts of each probe will also carry out some analysis of the ice composition at the Southern Pole.

Mars Polar Lander

The Mars Polar Lander will touch down less than 1000 km from the South Pole, near the edge of the carbon dioxide ice cap in Mars' late southern spring. The mission is designed to:

- record local weather conditions;
- analyse samples of the polar material for water and carbon dioxide;
- dig trenches and take photographs of the interior to look for layers that may have been put down from season to season;
- analyse soil samples for water, ice and minerals that may have been deposited by water;
- photograph the region of the landing site to look for evidence of climate changes and seasonal variations;
- attempt to determine soil types and composition of the local regolith.

The Lander will make a direct entry into Mars' atmosphere and will be slowed to a landing by a combination of aerobraking, parachute and descent rockets. It is due to last 90 days on the surface, although it is possible that it might survive longer.

Both of the following missions are currently under review due to problems developing the rover that was intended for the Lander and the project going over budget. Much of the information below is based on the original mission profile.

Mars Surveyor Orbiter 2001

The Mars Surveyor Orbiter is intended to act as a relay station for the Surveyor Lander as well as to conduct a detailed analysis of the mineral content of Mars' surface. It will also make measurements of the radiation levels. Launch is due to take place during a 20-day window from 7 March 2001. After a flight lasting nine months the probe should arrive at Mars, where it will use aerobraking to slow it into orbit. After some further manoeuvres, placing it into a 400 km circular orbit, it will open its solar panels and commence its survey mission. It is due to stop its survey after 1100 days, although it will continue to act as a communications link until the end of its projected five-year life.

Mars Surveyor Lander 2001

The Mars Surveyor Lander will be equipped to investigate the radiation levels at the surface, carry out experiments on the Martian regolith (including how toxic it might be to humans), and investigate the potential for manufacturing propellant on the surface. The Lander will also have cameras and a small robotic arm. The re-designed mission will use the *Marie Curie* rover, which will be almost identical in design to Pathfinder's *Sojourner* rover.

The Lander also has a 20-day launch window that opens on 5 April 2001. After a nine-month journey to Mars it will arrive between 16 January and 5 February 2002 (which should be something like 37 days after the Orbiter has arrived). This craft is designed to enter the atmosphere directly at something like 8 km s^{-1}, parachute part way down and land on braking rockets. The Lander should last 100 days on the surface, depending on the temperature at the precise landing spot.

After the 2001 missions, plans for the next two opportunities (2003 and 2005) are somewhat more fluid at the moment. There is every intention to use one of these slots to attempt a sample-return mission using a combination of robot sample probes and surface-manufactured propellant. The current very preliminary mission designs are as follows.

Mars Surveyor 2003

Probably launching in May/June 2003, this mission intends to use a long-range rover to search for evidence of life and organic materials. The Lander will have cameras, soil experiments and a small ascent vehicle. The rover will have a sample arm and a means of storing rock samples taken during its wanderings. The samples will be launched into Martian orbit by the ascent vehicle, where they will remain for collection at a later date.

Mars Surveyor 2005

Due for launch July/August 2005, this mission will fly along with a French-built orbiter that will collect samples from the 2005 Lander as well as those waiting in orbit from the 2003 mission. The Lander will be equipped to take samples from below the surface.

8.6 Mars in the more distant future[3]

With the aforementioned sequence of probes, NASA is laying the foundation for a manned landing. By the end of the 2005 missions there should be highly accurate maps of the surface to select landing sites from, extensive knowledge of the conditions on the surface, and samples of Martian rock returned to Earth for study. The first of the sequence of launches for a Mars Semi-Direct mission could take place when the next window opens in 2007. That will be thirty eight years after the first Apollo Moon landing.

Looking into the future beyond the first manned landing on Mars is an interesting exercise, but one that is difficult to do objectively as so much depends on motivation and economic viability. If manned exploration of the planet is to be of any use both scientifically and culturally, then it is clear that prolonged stays on the surface (measured in months, not days) are vital. Despite being a much smaller planet, the Martian surface represents a similar land area to that of the Earth. The potential for exploration is clearly vast, and the slightest indication on the surface that life may well have independently evolved there would trigger a veritable gold rush of scientists.

The only way that a scientific team can survive on the surface for reasonable periods is to develop a culture of 'living off the land'. Fortunately Mars is well suited to this sort of approach. Zubrin has already shown the way by indicating how Mars' natural resources could be used to generate propellant, but for a base to survive much more than that must be done. For some while, space enthusiasts have talked about the prospects of building a base on the Moon, but in fact Mars is a better prospect in that far less would have to be continually shipped from Earth in order to supply the base.

Power could be provided by a small nuclear reactor brought from Earth (provided the hazards of lifting the material from the surface are overcome) or by solar energy. Mars has a distinct advantage over the Moon as far as solar power is concerned. While the Moon's lack of atmosphere and comparative nearness to the sun means that the light is more intense, it also suffers from a 28 day 'day/night' cycle. The length of the Martian day is little different from that of the Earth—which helps compensate for the comparatively low intensity of light reaching the surface as far as generating power is concerned. Another advantage that Mars enjoys is that the materials for constructing solar cells (principally silicon, carbon and hydrogen) are far more abundant there than on the Moon. Geothermal[4] energy could also be employed to keep pace with incresing power demands.

There is abundant evidence visible on the surface of Mars to suggest that liquid water once flowed freely there. Photographs taken from orbit show dried up river channels and flood plains. Much of this water could still be present on Mars in the form of permafrost under the surface and ice at the polar ice caps. Estimations suggest that if all this water could be melted again (and the surface of the planet was completely smooth) then the liquid water produced would cover the entire planet with an ocean of depth 100 m. While recent discoveries[5] show that the Moon has far more water on its surface than had previously been thought, this is almost completely in the form of very solidly frozen ice in deep craters at the poles—where no sunlight can melt it. Given this a Martian base would be more easily able to provide its own water than one on the Moon.

Growing crops would also be far easier on Mars than on the Moon. Once again the less intense light on the Martian surface is compensated for by the short night periods compared to the Moon, where it would be necessary to illuminate greenhouses electrically at great expense. Furthermore, the lack of atmosphere on the Moon means that the greenhouses would have to use thick glass to protect the crops from the radiation and UV exposure that they would receive during a solar flare. Mars' atmosphere would provide sufficient natural protection. It should not be too difficult for a Martian base to become partially self-sufficient in terms of basic staples.

Looking to the even longer term, most space enthusiasts assume that at some point in the future a Martian colony will be set up. Undoubtedly the technology for this is a reasonable extrapolation from that we currently possess. Once a manned mission has been staged, then the process 'simply' needs to be repeated. The development of advanced propulsion systems such as those discussed in chapter 3 would make regular flights to Mars easier (and cheaper); however the real advance will come about with the development of cheap transportation into low Earth orbit. Much of the ΔV required to reach Mars comes from lifting the payload into Earth orbit, so it is in this part of the journey that the most significant cost savings have to be made. After all, systems such as the ion drive can only be used to provide a low thrust, not to lift from the ground.

Even with substantial cost savings it is difficult to see how governments could afford to fund a Martian colony, still less why they should want to. If a new society is to be set up on Mars, then it will have to be largely self-sufficient in terms of both resources and economics. The Martians must have something to sell.

They do. Mars is well stocked with a variety of minerals and other materials that will become increasingly rare on Earth in the future. Mining of them on Mars for transportation to Earth could become economically viable. However, it is likely that the most exportable resource that Mars has will be its abundant stocks of deuterium[6].

Modern nuclear reactors can only use enriched uranium fuel (or plutonium, which is expensive and highly dangerous)—which means that the proportion of the isotope $^{235}_{92}$U has been increased over that found in the natural ore. Such enrichment is very difficult, inefficient and costly. A reactor that uses heavy water—water in which the hydrogen is partially replaced by the deuterium isotope—can work with natural uranium, circumventing the need for enrichment. Deuterium is worth \$10 000 per kilogram. Once the technology for fusion reactors is finally developed, it is likely that this price will increase dramatically (the use of deuterium in the fusion process is covered in the next chapter). Deuterium is five times more abundant on Mars than it is on Earth.

Once a need is established it is easier to imagine a large company funding the transfer of people to Mars in order to develop the export of deuterium. As with so many other enterprises of this sort, it becomes much more practical to provide a living environment at the spot where the industry is developing than to ship people back and forth. This is probably how a Martian colony will start.

As the status of the colony improved, Zubrin has suggested that a triangle trade would be set up. Mars is very close to the asteroid belt (both in terms of distance and ΔV), which is composed of a very large number of rocky bodies that contain mineable natural resources such as iron, cobalt and nickel. Miners working in the asteroid belt would depend on Mars to supply them with food and probably oxygen. This would be a substantial export for the Martians. The asteroid belt would export metals to Earth, and Earth would most likely export high-technology goods to Mars.

While Zubrin is undoubtedly an optimist and is quite evangelical about Mars, his arguments hold weight. With the population of Earth growing continually and the demand for power and raw materials increasing in the developing world, I can see pressures for expansion of the human race into space growing. Mars is the most viable place for settlement. There is also another factor to consider. The Earth is becoming a smaller and more crowded place in terms of human interaction. If the prospect of emigration to Mars ever opens up, and with it the chance to live in wide open space and to develop a new society, I can easily imagine the attraction for a large number of people. The 'pioneer blood' that lead to colonization of new lands on Earth is undoubtedly still present in the human genetic make-up; it is just being suppressed at the moment as there are so few new places left to go.

Notes

[1] 'Report of the 90 day study on human exploration of the Moon and Mars.'

[2] This eventually became the International Space Station (ISS).

[3] Much of the information in this section is derived from Zubrin's book *The Case for Mars* which includes substantial sections on the colonization of Mars.

[4] This is energy extracted from the heat that is present under the ground.

[5] The recent *Clementine* and *Lunar Prospector* spacecraft have made measurements which suggest that small, frozen pockets of water ice may be embedded in shadowed regions of the lunar crust. Although the pockets are thought to be small, the overall amount of water might be quite significant, perhaps an amount the size of Lake Erie. This water may have originated in the comets that have continually bombarded the surface over the Moon's history. It may be trapped inside enormous craters—some 2240 km across and 13 km deep—at the lunar poles. Due to the very slight tilt of the Moon's axis compared with Earth's (1.5°), some of these deep craters never receive any light from the Sun.

[6] Deuterium, or heavy hydrogen, is an isotope of hydrogen consisting of one proton and one neutron.

Intermission 6

Godspeed John Glenn

This section was written an hour after watching the launch of the space shuttle carrying John Glenn into orbit for the second time. If the style here seems a little rough, that is because this section has been largely unedited so as to try to capture the feeling of the moment.

Thirty six years ago John Glenn became the first American to orbit the Earth. Today, 29 October 1998, he flew the space shuttle *Discovery* into orbit for his second flight (STS95). On his first flight he was sent on his way with the words of Scott Carpenter, the link man between the control room and the capsule (CAPCOM), resounding into history:

'Godspeed John Glenn'

The same words were used again today by the same man to the same astronaut.

After his first mission John Glenn returned to Earth a hero. The Soviets had placed a man into orbit before the Americans, but in a veil of secrecy. America was developing its space programme in the full glare of international publicity. John Glenn was the third of the Mercury astronauts but the first to take a ride into orbit. By all accounts he is a serious man not given to the practical joking that his fellow members of the Mercury Seven group were accustomed to. During his 4 hour 55 minute and 23 second flight he orbited the Earth three times in his tiny capsule *Friendship 7*. [Note that all the Mercury capsules had names ending with the numeral 7—this was in honour of the seven astronauts chosen to lead the American space programme.] As he passed over western Australia, all the lights were turned on in the city of Perth to welcome him. Apparently they will do the same again as he passes over in *Discovery*. During his first flight a fault indicator in mission control suggested that his heat shield might have come loose. As a result the ground controllers asked him not to jettison the retro-rocket pack that would normally be released once the rockets had fired to slow the craft into its re-entry corridor. They were worried that it might be

Figure I6.1 Contrasting launches—two ways of lifting John Glenn into orbit.

the only thing holding the heat shield in place. Initially they did not tell him. If it had been a Soviet mission, we would not have known about this problem. As it was, the world was captivated.

Glenn was intelligent enough to work out for himself that something was seriously wrong. Keeping the retro-rocket pack in place was a drastic change to the mission plans. As a result the autopilot that should have guided the capsule to re-entry had to be overruled. Glenn had to fly the re-entry himself. He admitted to a few anxious moments watching pieces of metal fly past the window. Fortunately it was only the retro-rocket pack burning up, not the heat shield failing. In the eyes of the world, John Glenn became a hero during those moments.

Reluctant to lose a genuine American hero to the dangers of space travel, President Kennedy grounded John Glenn after only one flight. Hence the thirty six year wait for the second trip into orbit.

In the meantime, Glenn turned his attention to politics, becoming Senator for Ohio in 1974. He unsuccessfully attempted to become the Democratic presidential nominee in 1984 and 1988.

At the age of 77, John Glenn's return to space is a curious amalgam. Sold to the American people as a genuine attempt to forward research into the ageing process (many of the reversible effects of space travel—e.g. thinning of the bone—take place naturally in later life), the mission is undoubtedly also a media boost for NASA and a nostalgia trip. There have also been suggestions that the flight was a political pay off from President Clinton to a Senator who had supported him through some troubled times. The latter is difficult to believe given the inherent dangers associated with space flight and the medical checks that the crew have to pass in order to be cleared. There is little doubt that Glenn would not fly unless he could hold his own with the rest of the crew.

300 000 people crammed the area around the launch site to see the shuttle take off—surely another piece of evidence that interest in the space programme is once again in the ascendancy. The President himself remarked in the run-up to the launch that this mission marked the end of one era of space flight and the start of another. Glenn's shuttle flight will be the last before the parts of the International Space Station start flying into orbit. Perhaps the point is that the space station will mark the start of the exploitation of space as an industrial and research environment rather than the high frontier that it has been so far.

However, the mission has a serious side to it as well. The effects of long-term exposure to zero *g* are well documented, but still not fully understood. The spine gets longer by about 2.5 cm, the heart rate slows and fluids in the body rise, making the face look rather fatter than it used to be. Also the various muscles that we use daily to support our own bodies against the pull of gravity lose definition. For this reason a regular routine of exercise is prescribed for lengthy stays in space. It is vital that these effects become better understood, both for the medical implications that such understanding may bring to the increasingly large proportion of our population over the age of sixty and also for the long-term future of mankind in space. A Mars mission would involve a prolonged stay in zero g. I am quite sure that the future of mankind lies in space. A permanent base on the Moon, a colony on Mars and mining operations in the asteroid belt seem to me to be inevitable features of our future. Perhaps STS95 will be the beginning of a new era, the era in which man at last started to take seriously the fact that eventually he will have to leave this planet and learn to live in space.

Chapter 9

Journeys to the stars

In this chapter we will look at the possibilities for constructing spacecraft and propulsion systems that are capable of travelling to the stars.

'Space is big. Really big. You just won't believe how vastly hugely mindbogglingy big it is. I mean you may think it's a long way down the road to the chemist, but that's just peanuts to space.'

From *The Hitchhiker's Guide To The Galaxy* by Douglas Adams

9.1 Motivations and timescales

In this chapter we shall look at some of the proposed ways of mounting manned expeditions outside our solar system. Some of the technologies are highly speculative; some are more realistic. In all cases they would involve a considerable commitment in terms of manpower and resources to develop and mount such an expedition. While it is quite possible that we might be able to send an unmanned probe to one of our nearest neighbours within the next 30 years, for the majority of the interstellar projects discussed in this chapter we are looking at timescales well into the next century, or possibly beyond that. With that in mind, there are several immediate questions that need answering.

1. What are the immediately achievable goals that we are likely to settle on in terms of space exploration in the next century?
2. Why are we bothering to go into space in the first place? Why would we want to mount an expedition to the stars? Why would governments wish to fund the immense costs involved?
3. Why do people seriously spend their time working on prospective technologies that have little chance of bearing fruit in their lifetimes?

Any attempts to answer the above questions must be based, to a certain extent, on a personal view. However, in trying to be as objective as possible I would suggest that the following are reasonable starts to furnishing answers, even if they may not be complete replies to the questions posed in everyone's view.

Point 1

Mars and the International Space Station (ISS) are going to be the focus of much of the space exploration effort in the early part of the next century.

Arguments for and against the ISS have been raging over the past few years, especially as America is having to carry an increasing burden of the projected costs since the Russian economy started having such spectacular problems. Some people have argued that the zero g environment makes certain types of medical and materials research much easier than would be possible on Earth. Others argue that we just have to be a little clever about the way in which it is done.

Aside from the construction of the ISS, developments in engine technology are likely to become more important as we move into the next millennium. The DS1 mission is already evaluating ion drive as a propulsion system and research into other techniques, such as those explored in chapter 3, is being given increased priority by NASA.

However, the most important issue by far is the development of a cheaper method for lifting payloads into low Earth orbit. This holds the key to the exploitation of space. In ΔV terms, once you have arrived at Earth orbit, the rest of the solar system is open to you. Without the cost of lifting cargo into orbit being significantly reduced the prospects of a sequence of manned Mars missions leading up to eventual colonization and trading between planets are not good.

Point 2

Exploration is a fundamental characteristic of the human spirit—a sentiment expressed frequently by the Apollo astronauts. This, coupled to the curiosity of the human mind with regard to the universe in which we find ourselves, will eventually express itself in the desire to study other star systems. Earlier in this book I quoted the words of Gus Grissom in commenting upon the relative merits of sending automatic probes and manned missions to the Moon. His words apply equally well to the idea of interstellar travel: '*Our God-given curiosity will force us to go there ourselves because in the final analysis only man can fully evaluate the Moon in terms understandable to other men*'. The potential for discovery and exploration increases dramatically as our range increases. While it is unlikely that any intelligent life will be found in the most immediately accessible star systems (otherwise radio communication would probably have been established by now), the possibility remains a great motivator. The implications that such a discovery would have for our ethical, religious and sociological views are explored fully in Paul Davies' excellent book '*Are we alone?*'.

Given all this there seems no doubt that should the opportunity arise there will be no lack of applicants for the first mission to the stars (me, for starters). However, there remains the question as to why governments would wish to finance such a mission. As I commented earlier, it is hard to argue against the feeling that the money spent on space travel would be better employed in medical research, feeding the poor, etc. Yet the fact remains that year on year governments do spend a fraction of their income funding pure research which has no obvious application in the short term. NASA and other bodies would, quite rightly, point to the spin-offs from space research that have enabled new technologies to be developed which have an immediate impact on people's way of life. We would not wish to place Velcro in this category, but a recent NASA publication listed an advanced wheelchair, a vehicle controller for the handicapped, water recycling, a device that converts regular inkprint into a readable vibrating form for the blind and a breathing system for fire fighters among the recent advances sourced from the space programme. In the end we have to recognize that pure research and exploration do raise the quality of life for everyone, not just through their spin-offs but also in the simple human achievements that they represent. Governments the world over clearly recognize this. However, this is not an argument that I would wish to push in conversation with a starving child in a poor country.

In any case, I believe that it is most likely that the funding for such expeditions would not come from governments but from private consortia. An idealistic view of the next century would envisage a society in which people had an enhanced fraction of their lives available for 'leisure' activities. If Mars and the asteroid belt do open up as a new frontier as many people suggest they might, then the populations that they support will become rich quickly as resources dwindle on the home planet. From this 'pioneer stock' will come the people with the desire and the money to mount such expeditions in collaboration.

Point 3

With many of the ideas outlined in this chapter, the people responsible for developing them harbour the optimistic hope that one might just turn out to be the key that opens up space travel for them.

The value of such speculation is that it quickly establishes what is and is not possible. It is very likely that a new idea not included in this chapter will eventually emerge as the breakthrough required, but as in any branch of research the key to progress is the ability to frame the right question. These ideas can be viewed as attempts to get the question right. Without this sort of speculation (beating the undergrowth to see what runs out) progress would never be made. NASA is starting a gentle attack on this issue with the instigation of its breakthrough propulsion physics (BPP) programme, more information about which can be found at the website: http://www.grc.nasa.gov/WWW/bpp/.

Finally, it must be remembered that designing systems that would enable travel to the stars is just plain fun.

9.2 The ground rules

The purpose of this chapter is to discuss methods that might be used to send a *manned* expedition to another star. Various possibilities come into play for an unmanned or robot probe, but such a mission would be a precursor for a more complex and demanding mission that carried a crew. Of necessity this chapter is based on speculation. Therefore it is important to understand the basis on which this speculation is being made.

If this were a work of science fiction, then we would be at liberty to create some modification to the laws of physics that would enable us to construct a propulsion system adequate to the task. However the game that we are playing here is a different one, and one that I think is ultimately more interesting. The rules of this game allow us to speculate freely about the advances that may be made in technology (both physical and biological), but to remain strictly within the laws of physics as we understand them at the moment.

Some science fiction works to similar rules and so has come up with useful contributions to the fund of ideas on how to go about a journey to the stars. I will mention some of these contributions in this chapter.

All of the propulsion systems that are outlined below are physically possible given an advance in technology (which may be an extreme advance in some cases). The engineering and physics of the systems have been worked out and demonstrated to function, in principle. Some of the ideas are more speculative than others, but they have all fuelled debate—which is the main purpose of the exercise.

The basic physical restrictions that we have to work around in order to send a starship exploring are as follows.

- Stars are a very long way away—the nearest star system to ours is Alpha Centauri[1] and it takes light (moving at three hundred million metres per second) 4.3 years to reach us from there (that is a distance of 4.1×10^{16} m).

- Einstein's special theory of relativity shows us that no objects can travel as fast as light—this theory has been very adequately checked by a number of experiments over some years and shows no sign of being wrong in any of its details;
- Travel times to even the nearest stars are likely to be significant fractions of a human life span. It is therefore unreasonable to transport adequate supplies with the ship—most likely food and water would have to be produced on board and atmospheric recycling would have to advance considerably over the current state of the art.

The latter point is very important as far as mission design is concerned (indeed it is the focus of the colony ship proposal below). One way of getting round the problem would be to employ a form of suspended animation, which may be plausible biologically. However, we would be unlikely to place the whole crew in suspended animation and trust to automated systems, so some supplies would have to be provided.

9.3 Is warp drive[2] possible?

No.

9.4 Orion and project Daedalus

In a sense I am cheating already in that the *Orion* spacecraft was never conceived as a means of getting to the stars. The idea behind Orion was to provide an effective means of transporting people and materials about the solar system. However, it did form the basis for a study carried out by the British Interplanetary Society which produced a design of starship—*The Daedalus*.

The principle on which Orion is based was first thought of by S Ulam and C J Everett, who wrote a paper on the subject in 1955. The idea was taken seriously by an Air Force committee, patented by the Atomic Energy Commission and developed by a team belonging to a branch of the General Dynamics Corporation in 1958.

The idea was to use atomic bombs to propel the spacecraft.

In evaluating this idea it is important to remember the time in which it was conceived. Sputnik had recently been launched and the American response was looking very weak by comparison. It was not that long ago that the world had been shocked by the detonation of the first atomic bombs and their deployment

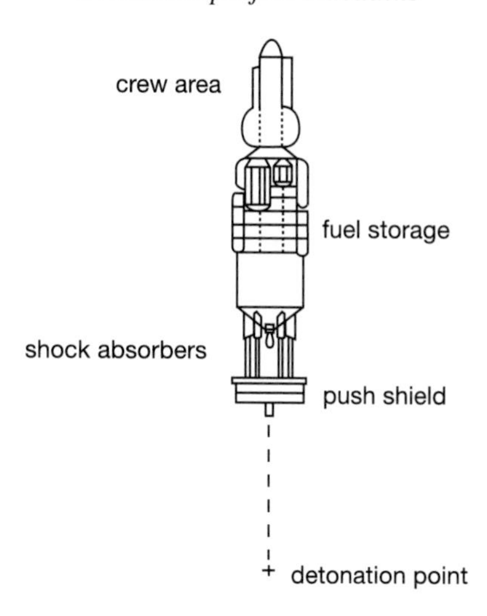

Figure 9.1 The Orion spacecraft powered by exploding atomic bombs.

against the Japanese during World War II. Many physicists, especially those who had been involved with the bomb's development, were horrified by what they had done. A peaceful use of such weapons and one that promised to open up the solar system to wide-scale exploration was a dream worth pursuing. The motto of the development team from the General Dynamics division (General Atomic) that took an interest in developing Orion was 'Saturn by 1970'.

Orion would basically be a large pusher plate attached to the main spacecraft by shock absorbers. The bombs would be dropped out of the back to explode beneath the pusher plate. The force of the explosion would impact on the plate and be translated via shock absorbers into a forward thrust on the spacecraft. Of course adequate shielding would be provided to protect the crew from the radiation and blast of the explosions.

The General Atomic team, lead by Theodore Taylor, carried out extensive tests using conventional explosives and models. After seven years of development the project was finally killed due to a variety of funding and political problems. The problems were not technical—the team had done the engineering very carefully and expertly. In the end the public perception of a craft powered in this manner was not an easy hurdle to mount. The ban on nuclear tests in Earth's atmosphere enforced in 1963 certainly did not help. One of the test models still exists and hangs in the National Air and Space Museum in Washington.

The physicist Freeman Dyson extrapolated the Orion spacecraft into a model that would be capable of interstellar travel. His version of the ship would have a payload of 20 000 tonnes (enough to support a crew of several hundred) with a total spacecraft mass of 400 000 tonnes—including 300 000 bombs of about 1 tonne each. The bombs would be detonated once every three seconds, accelerating the craft away from the Sun at a steady rate of $1g$ for ten days, by which time it would have reached a speed 1/30th of the speed of light. This would give a transit time of 130 years to Alpha Centauri—so it falls under the category of a colony ship as discussed later.

Daedalus was a design project undertaken by the British Interplanetary Society in the 1970s. The basis of the study was to send an unmanned probe to Barnard's star (about 6 light years away) with a trip time of 50 years.

The design was for a 49 000 tonne ship to be accelerated to 15% of the speed of light by using fusion explosions. Fusion is the merging of low-mass nuclei (such as hydrogen) into heavier ones (such as helium), and in the process producing energy. This sort of nuclear reaction provides energy in the core of a star (the central and hottest part of a star). Energy escaping from the core heats up the gases in the star, allowing them to resist the gravitational force tending to make the star collapse under its own weight. As long as the star is fusing nuclei in its core[3], it will remain stable—typically this state of affairs can last 8–10 billion years. The fusion process for hydrogen has three distinct steps:

1. proton + proton → deuterium + positron + neutrino
2. deuterium + proton → helium 3 + gamma ray
3. helium 3 + helium 3 → helium 4 + proton + proton

which produce a net energy release, mostly in the form of the kinetic energy of the products. This reaction is known as the proton–proton chain and is the commonest energy-producing reaction in typical stars such as our own Sun.

The deuterium nucleus is a proton and neutron bound together. Helium 3 is formed from two protons and a single neutron. Helium 4 is two protons and two neutrons together[4]. The Sun is able to carry out this combination of steps as it has very high temperatures, large numbers of particles and a great deal of time on its side.

High temperatures are required as only then have the protons enough energy to approach each other closely enough for a reaction to take place. Ordinarily protons repel each other due to their positive electrical charge. With enough energy the protons will be moving too fast for the electrostatic repulsion to produce enough of a momentum change to deflect them away from each other. Of course, we have to rely on chance for them to be approaching each other at all.

An individual proton would take part in the first-step reaction once every 10 million years or so. However, the Sun produces energy at the vast rate of 3.9×10^{26} W. This can only happen due to the huge number of protons available. An individual proton may take part in the reaction only very rarely, but with the number that the Sun has to play with the reaction is taking place at a regular rate. About 600 million tons of hydrogen are converted into helium *per second* in the Sun.

On Earth it would be impossible to generate energy by the same process—we could never bring the mass of hydrogen together to make it possible.

Various alternatives are being researched on Earth as a means of producing energy. Fusion would have the advantages of producing no radioactive waste and being easy to provide fuel for (you could use water, for example). However, as the masses involved are so much smaller than in the Sun, the temperatures have to be very much greater. Heating a gas to these sorts of temperatures and then containing it is proving to be a difficult problem.

Daedalus would use a direct reaction between helium 3 and deuterium. Fuel pellets consisting of a mixture of deuterium and helium 3 would be placed in a 'combustion' chamber where beams of high-energy electrons would heat the surface of the pellet and compress it at the same time, triggering the reaction. The result would be a stream of high-energy plasma, which could be directed by magnetic fields out of the probe, producing thrust. The fields would be produced by a large induction loop (shown in figure 9.2) which would have to be fed power from the main ship.

The craft would require 27 000 tonnes of helium 3, which is difficult to obtain on Earth. For this reason the mission design called for a stop at Jupiter where a series of robot probes would mine the giant planet's atmosphere for the isotopes required. The principles behind this mission also demonstrate that the Martian colonists may well have a market selling deuterium for use in space probes.

An alternative method for triggering fusion (also being explored for power stations on Earth) would be to heat the fuel pellets using a series of laser beams directed from different angles. In 1988 the US Navy and NASA produced a design (*Project Longshot*) for an interstellar probe to Alpha Centauri using fusion propulsion triggered by laser beams.

At the speed it would be travelling by the time it arrived at Barnard's star, Daedalus would only have a few days travelling through any solar system that the star might have. A variety of robot mini-probes would be deployed to survey the system.

Daedalus would also be equipped with an erosion shield designed to protect the critical systems of the craft. At the sort of speeds that Daedalus could achieve

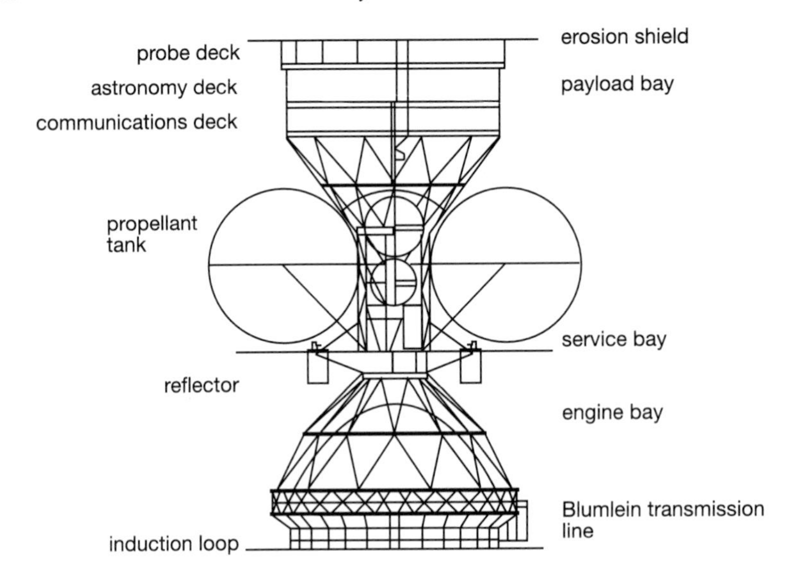

Figure 9.2 Project Daedalus.

even the smallest mote of dust would hit the ship with devastating force[5]. The erosion shield would be designed to steadily wear away under these impacts. In his novel *The Songs of Distant Earth*, Arthur C Clarke suggests using large blocks of ice keyed together as an erosion shield—ice would be easy to use and is quite tough enough for the job.

9.5 Laser propulsion

In chapter 4 I mentioned the possibility of using a large mirror to reflect sunlight and provide the thrust required for a powered geostationary orbit. Extending this idea we could use such mirrors to provide the thrust to propel a modestly sized spacecraft about the solar system. It would 'tack' on the light from the Sun rather as a sailing ship uses the wind. For this reason such a system is generally referred to as a *solar sail*. While this provides a very plausible method of propulsion within the solar system, it will not work for interstellar travel for the simple reason that sunlight decreases in intensity with distance. If the idea is going to work at all, then a beam of light of quite uniform intensity is required. Such a beam can be provided by a laser.

A laser is basically a source of light. However, unlike the light emitted from a bulb or fluorescent tube, laser light has properties that make it very useful for technological purposes. Primarily the light from a laser is coherent (meaning that all the waves making up the light are keeping pace with one another) and parallel. A reasonable laser should be able to produce a spot of light on a piece of paper that does not change in size by very much as the paper is moved away from the laser over a distance of several metres. This means that the light emerging from the laser is in the form of a beam, the edges of which are very close to being perfectly straight and parallel. With an ordinary lamp the rays of light are emerging travelling in all directions, exactly what is wanted for uniform illumination. Lasers are designed for concentrating light in very small regions—e.g. for a compact disc player where the laser is used to read the micron-sized pits and bumps used to record the music in digital form.

Robert Forward is one of the primary gurus of laser-propelled missions to the stars. In 1985 he proposed *Starwisp*—a very-low-mass (20 gram) interstellar probe constructed from a 1 km wire mesh sail with 4 grams of microcircuits embedded in the mesh. Such a construction is technologically beyond us at the moment.

Starwisp would be accelerated away from Earth at 115g to reach 1/5th of the speed of light in a few days. Such an enormous acceleration would be possible given the low mass of the structure. Acceleration would be by a beam of microwaves. The power required would be similar to that suggested for solar power stations in Earth orbit.

For many years scientists have been suggesting that one way of alleviating the demand for fossil fuels and other non-renewable energy supplies would be to construct solar power stations in Earth orbit. The continuous supply of sunlight unfiltered by the Earth's atmosphere would make them far more efficient than similarly sized stations on Earth. The energy could then be transmitted to the ground by a beam of microwaves.

Forward suggested using this beam, suitably modified by a giant microwave 'lens' constructed from wire mesh, to push the probe away. Once the desired speed had been reached the microwave beam could be turned off until the probe reached the target system.

Starwisp would then coast to Alpha Centauri, arriving 21 years later. On arrival the microwave beam would be turned on again at Earth to 'flood' the Alpha Centauri system with energy in order for the probe to collect and transmit a TV picture back to Earth. It has been estimated that, given a 10 GW microwave transmitter in Earth orbit, Starwisp might expect to pick up about 10 W of power at Alpha Centauri—which suggests a most efficient focusing of the power.

Using the same principle, Forward has also suggested a visible-light-powered starship capable of a return mission to Epsilon Eridani in 51 years total mission

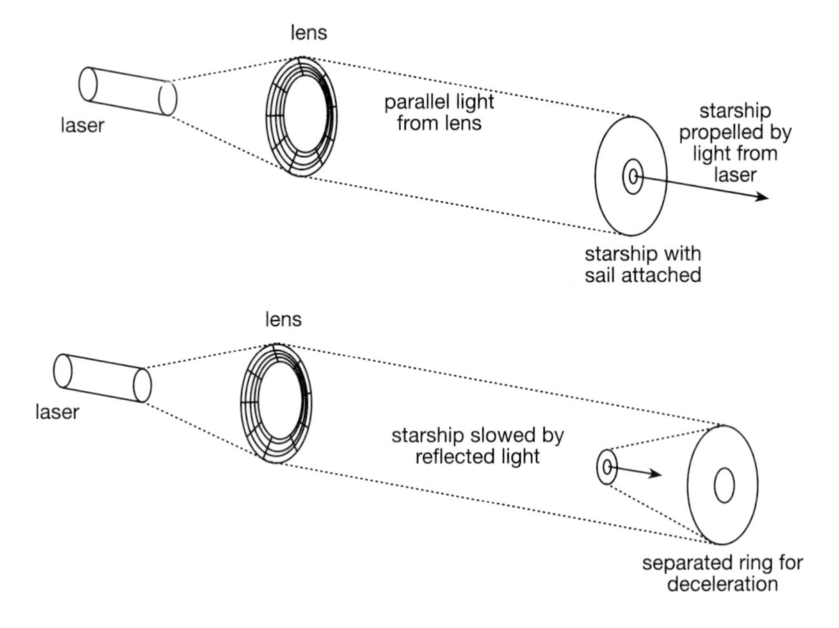

Figure 9.3 Robert Forward's scheme for pushing a starship using laser light.

time. The 80 000 tonne craft would have 3000 tonnes available for crew, supplies, exploration craft, etc. The rest of the mass would be a large (1000 km diameter) solar sail constructed from thin aluminium foil. Light to accelerate the craft would be provided by a giant laser in orbit around the planet Mercury (the nearest planet to the Sun in our solar system). The abundant sunlight in this vicinity would be used to power the laser, which would send out a beam to be collected and focused by a giant lens[6] in orbit between Saturn and Uranus.

The plan is to construct this lens from thin plastic film in a 1000 km diameter structure massing 560 000 tonnes. Such a lens would be capable of sending the laser beam over 40 light years before the beam spread out too much to be useful. The pressure of light from the laser would tend to distort the shape of the lens, which would need to be accommodated, and the scattering and absorption of energy along the length of the beam factored into the design.

Provided a laser capable of producing 4.3×10^{16} W of power could be constructed (which is an outrageous projection of current technology—the *total* power output the human race produces at the moment is 10^{12} W, which is forty thousand times less than required!), the spacecraft could be accelerated at $0.3g$, reaching half the speed of light in 1.6 years of continual acceleration.

Twenty years after launch the expedition would reach the target star and would be hoping to stop. The technique would then be to detach the outer portion of the solar sail, forming a large ring that, being lighter than the rest of the spacecraft, would accelerate ahead of it. This ring would also reflect light back to the ship which, as it had turned the remains of its mirror round to face the detached ring, would be decelerated.

The return journey could be achieved in a similar manner. The remaining mirror attached to the craft (the first ring that was detached would of course have been pushed well ahead of the target system) would once again be split. A section would be left behind in orbit of Epsilon Eridani (presuming the locals did not mind) to reflect the light from the laser back towards Earth. This reflected beam would be used to push the (now much lighter) ship on its return journey. Deceleration at Earth could be achieved simply by turning the remains of the mirror round so that it once again faced the laser, not the reflected beam coming from Epsilon Eridani.

Undoubtedly Forward's ideas are fraught with technological hurdles.

Just one example is the fact that in order to keep the laser beam focused on the starship the laser would have to keep its position in space accurate to a few metres! However, a search on the Internet shows to what extent Forward's ideas have influenced thinking, with people working on developing the basic system to make it a less extreme projection of current technology.

At the 46th International Astronautical Congress held at Oslo in 1995, Geoffrey Landis of the Ohio Aerospace Institute presented a paper[7] in which he considered how altering the laser wavelength and construction material of the solar sail might increase the workability of the design. He concludes that:

> 'Using a shorter wavelength, more efficient lasers, and a higher temperature and lower density berylium sail, the lens and the lightsail sizes required can both be reduced by a factor of five, the acceleration increased by a factor of twelve, the probe mass reduced by a factor of 33, the power requirement reduced by a factor of twelve, and the total energy reduced by a factor of 130 compared to the baseline mission. At today's electrical generating costs, the energy cost to launch the interstellar probe is only 6.6 billion dollars. This is quite reasonable for the magnitude of the mission proposed.'

In its turn this paper has caused controversy, with some people rejecting the principle outright. Others have suggested further refinements—for example using a beam of particles rather than electromagnetic waves. Whatever our views on the engineering of such a mission, Forward's ideas have clearly sparked discussion in just the manner that I suggested was the real value of this debate.

9.6 Ramjet

As we discussed earlier in chapter 3, one of the limitations to the performance of a propulsion system is the mass of propellant that has to be taken along as well as the payload. In 1960 R W Buzzard proposed a scheme that goes some way to solving this problem. He reasoned that there was enough propellant present in the vacuum of space to enable a ship to fly between stars.

Interstellar space is not completely empty. There are atoms of hydrogen and other elements and molecules spread thinly through space. Estimates of the density of such material vary, but it is likely to be clumpy with densities of the order of at least a few hundred atoms per cubic metre. Buzzard's idea was to use a large 'scoop' magnetic field to draw these atoms towards a starship, where they could be used as propellant.

The idea has several serious flaws.

Firstly, a magnetic scoop would only work if the atoms were actually charged ions, and it is not clear to what extent they are ionized. It might be possible to ionize them with an accompanying electrical field, but as the magnetic field would have to be several hundred kilometres across to scoop up enough material this would be difficult to say the least (it would represent a severe power drain on the ship if nothing else).

Another problem is that the scoop would only start to work efficiently once a reasonable proportion of the speed of light had been reached. Einstein's theory of relativity tells us that the density of material in space would appear to increase as the speed of the starship increased, allowing the scoop to work with more material. Consequently the ship would have to be equipped with an alternative drive system to get it going in the first place.

Finally there is the issue of what to do with the hydrogen that is scooped up.

Buzzard's original idea was to use the hydrogen in a fusion reactor, but that is problematical technology at the moment. Perhaps the material could be used as part of a matter/antimatter drive as outlined in the next section. Alternatively there are some schemes for combining the material that has been scooped up with stored propellant in a manner similar to that in which a jet engine works on Earth.

Buzzard's idea for gaining fuel from the material in interstellar space has merit, but it is an interesting adjunct to some other technology. It seems unlikely that it will be the answer to the propulsion problem on its own.

9.7 Antimatter drive

The first matter to establish with regard to antimatter is that it is not the province of science fiction. Antimatter exists. Indeed, a recent experiment at the main European nuclear and particle physics research establishment CERN (just outside Geneva) constructed anti-atoms of hydrogen for the first time. However, the experiment ran for three weeks, producing nine anti-atoms of hydrogen in that time, each of which lasted about forty billionths of a second—just about enough time to detect their presence, but little else!

Antimatter has been known about for sixty-odd years, ever since an experiment using cosmic rays produced evidence for a particle of the same mass as the electron, but with a positive charge. This is now known as the *positron*. Since then antimatter versions of other particles have been discovered and studied. The antihydrogen produced at CERN was constructed from an antiproton (same mass as the proton but with a negative charge) with a positron orbiting round it.

Some years ago, when antimatter was first being studied by physicists, science fiction authors took various liberties with its properties to produce convenient ways of moving between the stars. Many people assumed that as antimatter versions of charged particles have the opposite charge, they would have opposite gravitational effects. Useful though this might have been for propulsion, it has no basis in fact. Sensitive experiments have shown that antimatter behaves just like matter under the influence of gravitational forces.

However, there is one popularly known property of antimatter that is well established in fact. This is the tendency for matter and antimatter to annihilate one another into energy. This provides a useful basis for driving a starship.

When fundamental particles such as the electron (e^-) annihilate with their antimatter partners, in this case the positron (e^+), the result is a burst of electromagnetic radiation. However, when an annihilation takes place between more complex objects, such as a proton[8] (p) and an antiproton ($\bar{p}$), then the reaction is also more complex:

$$\begin{aligned} \text{simple annihilation}: &\quad e^- + e^+ \rightarrow \text{gamma rays} \\ \text{complex annihilation}: &\quad p + \bar{p} \rightarrow \pi^+ + \pi^- + \pi^0. \end{aligned}$$

Without going into the details of how such reactions work, the important aspect of this is that the complex reaction produces a string of particles, some of which (the π^+ and the π^-) are electrically charged. In such a reaction the energy is partially released as the kinetic energy of the produced particles (some of the energy goes into creating the particles even if they are stationary). This energy can be harnessed.

When a reaction such as the annihilation between a proton and an antiproton takes place in a material, the particles produced fly off through the material. As they pass through they are bound to collide with the atoms within the material. When this happens some of the particles' energy is transferred to the atoms. This makes the atoms vibrate about their average positions (if the material is a solid) or move about more freely (if the material is a liquid or gas). In other words the material heats up. This heat can be extracted as a useful form of energy. The amount of energy release from a single event is pitifully small on a technological scale, but in terms of the mass of reacting particles that has to be stored, a matter/antimatter reactor producing many such reactions a second is a very efficient source of energy.

The heat energy extracted from a matter/antimatter annihilation reactor could be used in two ways:

- Direct drive: heating a propellant mass to eject out of the craft in order to provide thrust. This would be very similar to the nuclear engines discussed in chapter 3, but the matter/antimatter reactor would be far more efficient than a conventional nuclear reactor.
- Power production: using the heat to generate electricity (in a similar manner to a conventional power station) to provide power for an ion drive.

Either of these techniques seems promising technologically. The decision between them would be based on the specific application—the direct drive would produce more thrust, but the ion-drive-powered matter/antimatter reactor would achieve a greater ΔV overall due to the faster exhaust speed achievable with ions.

The catch in all this is that the technology for manufacturing and storing antimatter is in its infancy. Currently we would not be able to produce anything like enough antiprotons to power a craft, let alone store them long term. However, this is certainly one of the more promising lines of research currently being explored.

9.8 Colony ships

This is not so much a propulsion design as a mission philosophy. Even with the most optimistic projection of technology mentioned above, it is likely that interstellar journeys will take longer than a reasonable fraction of a human lifespan. Unless some suspended animation technology can be used, it is probable that the crew that set out on a mission to anywhere other than the nearest stars would not survive to see the end of the journey. Under these circumstances the only option is to send a huge craft with adequate room and facilities for the crew to live comparatively normal lives, have families and train the next generation of explorers.

Such a ship would obviously have to be vast. It would need to be big for two reasons:

- a large biosphere is required in order to achieve stability;
- a large population is needed in order that genetic diversity (and free choice of mate!) be maintained—an important factor in ensuring the colony's survival.

The simplest design would be cylindrical. Inside the cylinder the inner walls would be lined with soil to a sufficient depth to allow trees and other plants to flourish. There would have to be room for free-standing and running water to be incorporated, as well as a means of illumination. Many writers have suggested a plasma tube down the axis of the cylinder. As it is unlikely that the whole cylinder would be filled with atmosphere, the plasma would be in a vacuum and so could be contained by magnetic fields. A current running through the plasma would cause it to glow, providing illumination—just like a giant fluorescent tube.

Taking up the point about the atmosphere, it would be unnecessary to fill the whole chamber with an oxygen/nitrogen/CO_2 mix (one could not maintain a biosystem in the same way as an Apollo command module—just on pure oxygen). As long as a sufficient depth of atmosphere was maintained above the curved walls this would have the advantage of leaving a partial vacuum down the centre through which craft could move without the impediment of air resistance. The ship would have to be spinning in order to simulate gravity and this would help keep the atmosphere in place. The spin would also enable a day/night balance to be maintained. Large curved plates next to the plasma tube could be used to shade half of the 'floor'. If these plates were on bearings at each end of the tube then as the ship rotated about the central axis the region of the floor in shadow would change—simulating night and day.

At one extreme we could be talking about a colony ship similar is size to the space station conceived in the science fiction series *Babylon 5*. The Babylon station is projected to be five miles in length and home to a quarter of a million people. With a mass of 1.5 million tonnes we are certainly talking about large-scale engineering.

Gerard O'Neil has considered the construction of a space colony in permanent orbit in our solar system, a design that could easily be adapted to a colony ship powered by an antimatter drive. His calculations suggest that a crew of 10 000 would require 45 m^2 of land area per person and 0.64 km^2 of intensive growing area. To construct this would require 100 000 tonnes for the structural mass, several hundred thousand tonnes for the buildings, soil and atmosphere, and some millions of tonnes in radiation shielding.

For current engineers the whole construction seems daunting—never mind the financial implications. O'Neil has suggested that a colony space station of this sort of design could export power generated by solar panels as a means of repaying initial investment. Perhaps, after a suitable period of doing this, the colony could be converted into a giant starship.

9.9 Wormholes

It is impossible to do justice to the subject of *wormholes* without a book in its own right. Wormholes are an unexpected solution to the equations of Einstein's theory of general relativity (the theory that refines Newton's law of gravity when the gravitational field is very strong). While related to black holes (and wormholes possibly form when two black holes in different parts of the universe merge in hyperspace), wormholes are different. The equations suggest that two wormholes in different parts of the universe could be connected by a 'corridor' through hyperspace. Travelling along this corridor would enable us to move from one wormhole to the other—a journey that might be considerably less distance through the corridor than via the more normal route through space. It is a little like travelling from England to New Zealand by going through the planet rather than round the surface.

Einstein's theory compares gravity to the shape of space. Admittedly it is very difficult to understand how space (which most people think of as nothingness) can have shape. In Einstein's terms, the fact that all objects fall with the same acceleration in a gravitational field no matter what their mass indicates something about the nature of the space through which they are moving. The standard analogy is to think of a stretched rubber sheet. A heavy ball placed on the sheet simulates a planet. Near to the ball the sheet deforms in order to support the ball's weight. This deformation takes the form of a bowl-shaped depression in the sheet. Now, if we were to roll another ball across the sheet, its path would be deflected by the depression. It would not be that the large ball exerted some sort of force on the smaller one, rather the space through which the small ball was travelling had been distorted by the presence of the large ball.

Extending this idea to our solar system and the planets surrounding the Sun is very difficult, even to professional physicists. If it were easy to imagine what the equations are telling us then we would not have to resort to analogies like the rubber sheet. Suffice to say that the equations of Einstein's theory tell us that the space (and time) in the vicinity of the Sun (or any other large mass

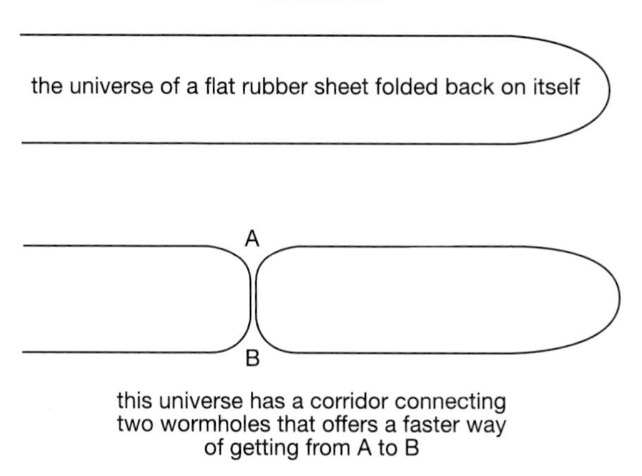

Figure 9.4 Cross sections through two rubber-sheet universes. In the second one two wormholes have linked different parts of the universe by forcing a corridor through hyperspace.

for that matter) is very different to that in deep space—away from any large masses. In deep space it is like a flat rubber sheet—spacecraft can move in a straight line at constant speed (as Newton's first law tells us). Near to a star, the spacecraft still moves in the straightest line that it can—through curved space.

Now the analogy of the rubber sheet can be as confusing as it is helpful. It is the surface of the sheet that is supposed to represent our universe. That surface is a two-dimensional object, yet the balls placed on it are three-dimensional. In the real universe the space is three-dimensional, as are the objects that are distorting it. In the case of the rubber sheet, the depression is a curvature 'into' the third dimension. Extending this into the universe, the curvature must extend into a higher dimension—one that is not part of the universe that we naturally inhabit. It is this extra dimension (or possibly more than one) that some physicists refer to as *hyperspace*. If we are to follow the analogy correctly, then we must imagine a race of flat creatures sliding about the surface of our rubber sheet. Their universe is that two-dimensional plane. They cannot leave it and enter the third dimension. Yet they can experience its existence indirectly.

Imagine a flat mathematician on a voyage of discovery. While sliding along the flat part of the sheet she notes that she can move in a straight line at constant speed without ever meeting an obstruction. Yet on a later voyage, that happens to pass near to the large sphere, she notes that moving in a straight line at constant speed has brought her back to her starting point (round the lip of the depression). She certainly moved in a straight line as no corners were turned. She has learnt something about the three-dimensional geometry of her world without leaving the surface.

Similarly, we can learn about the geometry of our universe without ever leaving it.

Now, to relate this to wormholes we must imagine a different rubber sheet, one that has been folded over onto itself. A wormhole would be a junction in which the curvature of space is so great that it has bent the real universe into hyperspace[9]. We can see from figure 9.4 that moving through the corridor would be a much shorter journey than sliding around the outside.

So, given what has been said about wormholes it would appear that there are reasons for optimism regarding the possibilities of travelling between the stars.

Unfortunately not. Several snags have arisen in the developing theory of wormholes:

- If they exist at all—and just because they are a solution to a set of equations this does not force them to have physical reality—then they are likely to be sub-atomic in size. Hardly encouraging for space travel.
- It may be possible to open the throat of a wormhole to a size large enough to permit travel, but only at great expense in terms of energy and by using a form of matter that may not exist in the universe—and one that we have no idea how to make.
- It is possible that wormholes may destroy themselves by magnifying energy that passes through them.

Wormholes are proving to be an entertaining way in which physicists can test their understanding of Einstein's theory. It is possible that an advanced civilization could use them to travel between the stars, but the poorly understood laws of physics in this area could equally well forbid them from existing at all.

9.10 Personal conclusions

Given the laws of physics as we currently understand them, the prospects for travel to the stars seem limited and rather constraining.

I am sometimes asked why we believe so strongly that faster-than-light travel is impossible when so many previous scientific ideas (that we can never travel faster than the speed of sound, for example) have been shown to be wrong. The reason is to do with the nature of scientific progress.

All the experiments that we have ever performed have shown us that the fundamental limitation on the speed of travel is correct. It is quite possible that in the future some experiment will show that some of the details in Einstein's theory are incorrect, but even then the basic premises will remain true. In the same manner, when Einstein discovered the general theory of relativity, it didn't significantly alter the results that we can obtain using Newton's laws on the scales that we experience gravity.

Nevertheless, it is possible that some new discovery might show the way to circumvent Einstein's constraint. However, I cannot speculate as to what this might be—so it seems to me to be profitless to discuss extra-solar exploration on the basis of a total unknown.

A different approach to the concept of exploration is required.

Timothy Ferris has come up with an intriguing possibility[10]. Given that travel between star systems is likely to be time consuming and expensive, and that radio communication is also a lengthy process, he suggests that a group of civilizations would most likely cooperate on constructing a network of relay stations. The network would function in a very similar way to that in which our Internet is set up. Automated systems placed in strategic positions about the galaxy would collect radio traffic and route it to its intended destination. Even with this system in operation, conversations between civilizations would be impractical due to the time delays involved. The suggested solution is also modelled on our Internet—what civilizations generally want from each other is information, not conversation. If each race were to set up a 'home page' with information regarding themselves and their star system, then other races could quite happily explore the galaxy from the comfort of their own planets.

Now, this nice scenario does skate over the formidable technical difficulties involved in such an enterprise. Simply translating from one language to another can be insuperably difficult on the same planet[11], never mind between the languages of biologically different creatures. Not only that, the technology is likely to be incompatible as well (unlike in the film *Independence Day*, it is probably not very easy to get an Apple PowerBook to talk to an alien computer in real life . . .). However, Ferris' vision is attractive and who can say what degree of cooperation and technology an advanced civilization would be capable of.

Even given this model of universal couch potatoes tuning in to the information provided by others, it is more than possible that the human race will wish to send representatives to the stars.

Given that there is little prospect of reporting results and discoveries back to Earth, we have to rethink the basic motivations for such exploration when it extends beyond our own solar system. Having spent so much time in travel, the eventual arrivals at a new star might reasonably think that there was no reason to report back to the people that sent them out. What allegiance would they have to Earth?

Even if they sent a message reporting arrival it would not arrive for at least several years—perhaps decades. Given that, it is equally valid to ask what allegiance the people on Earth would feel to the expedition?

Under such conditions it appears to me that the most sensible use of resources would be to not send people at all. I would send DNA and grow the people when the ship arrived.

This suggestion generally brings horrified reactions from anyone who hears it. Yet I believe that this is based on an inappropriate concept of exploration for the circumstances. People argue that it would be an inhuman thing to do to the 'people' sent. Why? I am not for one moment suggesting that they would be a 'slave race' sent to explore at the whim of Earth and to report back. My interest is in the long-term preservation of the human race and its diverse cultures. I would not want them to explore for us (for themselves, yes)—I would want them to colonize.

Why do we wish to have a space programme in the first place? I discussed this earlier in this chapter and suggested that the basic motivation is to satisfy the human thirst for exploration and knowledge, but that a useful side effect was the benefit that this had for technology. However, in taking a far more long-term view another issue raises itself: ensuring the survival of the human race. One thing is certain—eventually the Sun will die and then not just the Earth but the whole solar system will be uninhabitable.

Now I accept that it is very difficult to project into the future that far. The Sun is currently enjoying a comfortable middle age that is likely to last about four billion years. By the time that we will have to evacuate this star system it is quite likely that the 'human' race will not resemble us in any details of physiology or psychology. However, the natural resources of Mars, the asteroids and other planets will almost certainly not last as long as the Sun, and the pressures of population growth coupled with the exploratory urge will drive people to the stars eventually. As I have argued, aside from to the very nearest stars such missions cannot really be mounted in any way that allows exchange of information on a reasonable time scale—and certainly not the return to Earth of the same people that left.

I imagine a scenario very similar to that already suggested by a number of authors[12]. A fleet of comparatively small craft (certainly nothing like as large as a colony ship), equipped with some form of artificial intelligence, sent out into space. Possibly they would be programmed to use the resources that they came across to build other versions of themselves (then they would reproduce like a virus and spread across the galaxy more rapidly[13]). When they found hospitable planets they would land and set about preparing a colony for the first humans that they raised from the DNA loaded on board.

Given some very simple assumptions it is amazing what can be achieved. If D_{av} is the average distance between stars that could provide the raw materials for building probes, v the average speed at which the probes travel and T the time required for a probe to build a copy of itself, then the approximate time between one probe leaving a system and producing a second probe would be:

$$\frac{D_{av}}{v} + T.$$

Taking some conservative assumptions (such as $D_{av} = 5$ light years, $v = 91\ \mathrm{km\ s^{-1}}$ and $T = 100$ years) then this produces an expansion rate of 3×10^{-4} light years every year[14]. At this rate the whole galaxy could be covered in about 300 million years.

Now, it is difficult to judge how much credence to give ‘back of the envelope’ calculations such as this. For one thing, it does not account for natural disasters that might destroy a probe or render it incapable of reproducing. Nor does it take into account the intelligent direction of the probes (i.e. sending them towards likely target stars and not to stars that already have a probe on the way)—it assumes a random dispersal. The real value of the calculation is that it demonstrates that the idea of colonization in this manner is not obviously fundamentally flawed.

Expansion across the galaxy is plausible in a timescale very much smaller than the lifetime of the galaxy itself.

It makes one wonder if it has already been done.

Notes

[1] The Alpha Centauri system is actually a triple star system that contains Proxima Centauri, which many people mistakenly believe to be nearer the Sun than Alpha Centauri.

[2] As conceived by the writers, warp drive distorts space round a craft which then rides this distortion in a similar manner to that in which a surfer rides a wave. I am a great *Star Trek* fan (especially *The Next Generation*), but much of the science is not realistic.

[3] It is possible for fusion to take place in layers surrounding the core in high-mass stars later in their life.

[4] In the first step the positron is produced moving at very high speed. The neutrino is a massless (probably!) particle with no electrical charge. These neutrinos can be detected coming from the Sun in vast numbers. However for some reason that has not yet been adequately explained, the numbers of neutrinos produced is somewhat less than the theory would suggest.

[5] As a matter of interest, the problem would be far worse if it were possible to travel faster than light. This is the reason why all the starships in *Star Trek* carry deflectors (not to be confused with shields). It is the job of the deflector to sweep space ahead of the ship clear of any debris that could cause trouble.

[6] This would not be a lens similar to the shaped pieces of glass found in spectacles. The design would be for a zoned plate constructed from alternate concentric rings of plastic and empty rings. With the correct sort of plastic this sort of structure can focus just like a conventional lens.

[7] Landis G 'Small laser-propelled interstellar probe' Paper IAA-95-IAA.4.1.102, *40th IAF Congress (Oslo, Norway, October 1995)*. For a reference to this paper on the Internet, see the appendix at the end of this book.

[8] Protons are not fundamental particles. We now know that objects such as the proton and neutron are composites of smaller elementary particles called quarks.

[9] Note that in travelling down the corridor, one is not travelling through hyperspace. You are still in our universe, just in a part of it that has been very seriously bent. In England we have a name for this—it is called going through Milton Keynes.

[10] Ferris T 1988 *Coming of Age in the Milky Way* (London: Bodley Head)

[11] Without the Rosetta stone, Egyptian hieroglyphics would probably never have been translated.

[12] For a fuller discussion of the issues in this section, as well as references to various papers and articles on the subject, see *The Anthropic Cosmological Principle* by John D Barrow and Frank J Tipler.

[13] Such a space probe is an example of a *Von Neumann machine* (after the mathematician who first considered the general principles of such a machine). Von Neumann machines are machines that are given two basic instructions—carry out some task and build a copy of oneself.

[14] These are not unreasonable assumptions given the distribution of reasonable stars and that 91 km s^{-1} is just greater than the speed needed to escape the solar system (which has already been achieved by Pioneer and Voyager space probes). T is open to some debate. However, given that it took 300 years to build the USA up to an industrialized nation from no basis, a sufficiently well programmed space probe ought to be able to develop a colony to the point of sending another probe out in the same order of time.

Appendix 1

Glossary

Aphelion
The point on an orbit about the Sun where the distance to the centre of the Sun is the greatest.

Apocynthion
As with apogee, but for an orbit about the Moon.

Apogee
The point on an orbit about the Earth where the distance to the centre of the Earth is the greatest.

Apolune
The point on an orbit about the Moon where the distance to the centre of the Moon is the greatest.

Astronomical unit
A unit of distance used by astronomers, the AU corresponds to the mean distance between the Earth and the Sun, 1.496×10^{11} m.

Avionics
The combination of aviation and electronics.

Booster
A general term used variously to mean the first stage of a multi-stage rocket, or the whole rocket without the payload.

In this book I have used the term in the sense of the whole rocket.

Cape Canaveral
The military testing and missile facility next to the Kennedy Space Center on the coast of Florida.

Cape Kennedy
No such place. See Cape Canaveral or Kennedy Space Center.

Composite
A type of material that is a combination of two or more other materials. For example, fibreglass is a combination of glass fibres and a resin to bond the material together. Composites are chosen to marry the properties of the materials in order to make a single unit that is better suited to a task than the individual materials would be.

Conjunction
Any two celestial bodies are in conjunction when a third reference body lies on an extension of the line joining the two bodies. The term most commonly refers to the position of the Sun and one of the outer planets as observed from Earth. See also *opposition*.

Cryogenic
Ultra-low temperature. Cryogenic liquids need to be stored at these low temperatures or they will boil into gas. Liquid hydrogen and liquid oxygen are very commonly used cryogenic liquids in space flight. They need to be stored at $-250\ ^{\circ}C$ and $-180\ ^{\circ}C$ respectively.

Electrolyte
A liquid through which electricity can be conducted. Normally the liquid parts of a wet cell battery or fuel cell. Electrolytes conduct electricity as they contain ions in solution.

EVA
Extravehicular activity—any activity that takes place outside the spacecraft in space. An *EVA suit* is another term for a space suit.

Fuel
Part of the propellant used to power a chemical rocket motor. The fuel is used in a chemical reaction with the oxidizer to produce a high-temperature exhaust.

Golf
A game which involves propelling a small white pellet into a small round hole from a great distance by using implements designed to make the task as difficult as possible.

Ion
The name given to an atom that has an extra electron added to it (and so is negatively charged) or an electron removed from it (and so is positively charged).

Kilogram (kg)
The SI unit of mass. 1 kg is 2.2 lb.

Kennedy Space Center
The NASA launch facility on the coast of Florida. Here the Apollo flights were launched and the space shuttle now takes off and lands.

Laser
A device for producing a concentrated beam of light that is *coherent* (i.e. all the waves in the beam are in phase) and nearly *parallel* (i.e. the beam diverges only due to diffraction). The word is an acronym formed from light amplification by stimulated emission of radiation.

Litre
The SI unit of volume. 1 litre is 0.26 US gallons.

Mach number
Mach number is the ratio between the speed of an object moving through a fluid and the speed of sound in that fluid. A Mach number greater than 1 indicates that the object is moving at a hypersonic velocity. The speed of sound in air at sea level and 15 °C is 340 m s^{-1}.

Metre (m)
The SI unit of length. 1 m is 3.28 ft.

Newton (N)
The SI unit of force. Engine thrust is often quoted in pounds. To convert this into newtons the correct number of kilograms needs to be calculated and then multiplied by 9.8 N kg^{-1} (the strength of gravity on the Earth's surface) to give the force. A 1 lb force is 4.4 N.

Opposition
Any two celestial bodies are in opposition with respect to a third when they lie on diametrically opposite sides of the third body. See also *conjunction*.

Oxidizer
Part of the propellant used to power a chemical rocket motor. The oxidizer is needed to react with the fuel in a chemical reaction similar to the way in which substances burn in the atmosphere.

Pascal (Pa)
The SI unit of pressure corresponding to a 1 N force over an area of 1 m^2. Normal atmospheric pressure at sea level is about 100 kPa (or about 15 lb per square inch).

Pericynthion
As with perigee, but for an orbit about the Moon.

Perigee
The point on an orbit about the Earth where the distance to the centre of the Earth is the smallest.

Perihelion
As with perigee, but for an orbit about the Sun.

Plasma
A plasma is a very high temperature gas in which each atom, or molecule, has been electrically charged by having electrons removed. At low temperatures the electrons would simply bind with the atoms or molecules again.

Propellant
The combination of fuel and oxidizer used to propel a chemical rocket. When the two components are mixed a chemical reaction takes place, producing high temperatures and pressures in the exhaust gas. In more advanced engine designs the propellant is a single liquid, such as liquid hydrogen, that is either heated to boil and escape from the engine, or alternatively is ionized and accelerated out as a stream of charged particles.

Sintering
The process by which a material containing a fibre (or powder) is heated to less than the melting point of the fibre and compressed to fuse the fibres together.

Specific impulse
A quantity used to rate the performance of various propellants. Specific impulse (measured in seconds) is the thrust produced per unit weight of propellant consumed per second. This is equivalent to the time needed to burn one kilogram of propellant in order to produce one newton of thrust.

Telemetry
The continuous stream of information of ship systems, course and thrust relayed to ground from a spacecraft.

Thrust
The rocket engineer's term for the force produced by an engine. Thrust can be calculated by multiplying the speed at which the exhaust leaves the engine by the rate at which propellant is being consumed. The result is measured in newtons.

Von Neumann machine
A machine designed to perform a specific set of tasks, one of which is to construct a copy of itself. Such a machine would, given the appropriate materials, reproduce like a virus.

Appendix 2

Apollo mission summary

Mission name	Crew (in order MC/CMP/LMP)	Spacecraft names (CM/LM)	Comments
Apollo 1 27 Jan 1967	Lt Col Virgil I Grissom (USAF) Lt Col Edward H White (USAF) Lt Comdr Roger Chaffee (Navy)	—	Fire during ground test took lives of all three men. Posthumously designated Apollo 1
Apollo 4 4 Nov 1967	No crew	—	First flight of Saturn V. Unmanned command and service module placed in orbit
Apollo 5 22 Jan 1968	No crew	—	Unmanned flight test of lunar module
Apollo 6 4 April 1968	No crew	—	Unmanned test of Apollo
Apollo 7 11–22 Oct 1968	Capt Walter M Schirra (USAF) Maj Donn Eisele (USAF) Walter Cunningham	—	Saturn 1B launch. 163 orbits of Earth in command and service module
Apollo 8 21–27 Dec 1968	Col Frank Borman (USAF) Capt James A Lovell (Navy) Lt Col William Anders (USAF)	—	First manned Saturn V launch. Flew to Moon and orbited for 10 orbits
Apollo 9 3–13 March 1969	Col James A McDivitt (USAF) Col David R Scott (USAF) Russell L Schweickart	*Gumdrop* *Spider*	First manned flight of lunar module in Earth orbit. Two EVAs. 151 orbits
Apollo 10 18–26 May 1969	Col Thomas P Stafford (USAF) Comdr John W Young (Navy) Comdr Eugene E Cernan (Navy)	*Charlie Brown* *Snoopy*	31 orbits of the Moon. Lunar module descended to within 9 miles of the surface
Apollo 11 16–24 July 1969	Neil Armstrong Lt Col Michael Collins (USAF) Col Edwin E Aldrin (USAF)	*Columbia* *Eagle*	First lunar landing. One EVA lasting 2 hours 48 min. Collected 48 lb of samples

Apollo 12 14–24 Nov 1969	Comdr Charles Conrad (Navy) Comdr Richard F Gordon (Navy) Comdr Alan L Bean (Navy)	*Yankee Clipper* *Intrepid*	Landed Ocean of Storms, close to Surveyor 3. Two EVAs, 7 hours 46 min total. Collected 75 lb of samples
Apollo 13 11–17 April 1970	Capt James A Lovell (Navy) John L Swigert Fred W Haise	*Odyssey* *Aquarius*	Landing aborted after explosion on service module put crew in serious risk. Lunar module used as 'lifeboat' to keep crew alive until re-entry
Apollo 14 31 Jan–9 Feb 1971	Capt Alan B Shepard (Navy) Maj Stuart A Roosa (USAF) Comdr Edgar D Mitchell (Navy)	*Kitty Hawk* *Antares*	Landed at Fra Mauro. Two EVAs, 9 hours 23 min total. Collected 94 lb of samples
Apollo 15 26 July–7 Aug 1971	Col David R Scott (USAF) Maj Alfred M Worden (USAF) Lt Col James B Irwin (USAF)	*Endeavour* *Falcon*	Landed at Hadley Apennine. Three EVAs, 18 hours 46 min total. Collected 169 lb of samples. First use of lunar rover
Apollo 16 16–27 April 1972	Capt John W Young (Navy) Lt Comdr Thomas K Mattingley (Navy) Lt Col Charles M Duke (USAF)	*Casper* *Orion*	Landed at Descartes Highlands. Three EVAs, 20 hours 14 min total. Collected 213 lb of samples
Apollo 17 7–19 Dec 1972	Capt Eugene A Cernan (Navy) Comdr Ronald E Evans (Navy) Dr Harrison H Schmitt	*America* *Challenger*	Landed at Taurus-Littrow. Three EVAs, 22 hours 4 min total. Collected 243 lb of samples

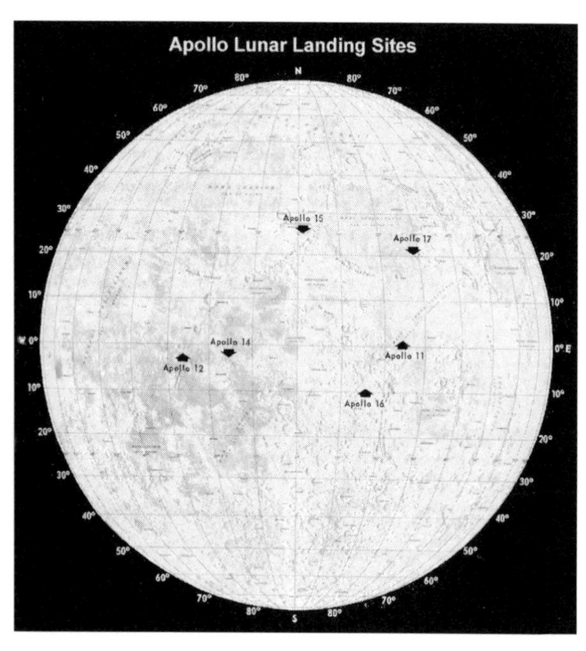

Appendix 3

Development of boosters

Apollo launch vehicles

Only those rockets with a '**' after their names were developed and flown.

Saturn C-1 (renamed Saturn I)**

First stage	S-I booster	eight H-1 engines; total thrust 6.7 million newtons
Second stage	S-IV	four engines using LH_2/LOX; total thrust 355 800 N
Third stage	S-V	two engines like those in the S-IV stage; total thrust 177 900 N

In March 1961, NASA approved a change in the S-IV stage to six engines that, though less powerful individually, delivered 400 300 N (90 000 lb thrust) collectively. On 1 June 1961, the S-V was dropped from the configuration.

Saturn C-1B (renamed Saturn IB)**

First stage	S-IB	eight modified H-1 engines; total thrust 7.1 million newtons
Second stage	S-IVB	one J-2 engine; total thrust 889 600 N

This booster was used to lift both manned and unmanned Apollo spacecraft into orbit.

Saturn C-2

Four-stage version:
S-I booster
S-II second stage (not defined)
S-IV third stage
S-V fourth stage.

Three-stage version:
S-I booster
S-II second stage (not defined)
S-IV third stage.

Plans for the C-2 were cancelled in June 1961 in favour of the C-3.

Saturn C-3

First stage	two F-1 engines; total thrust 13.3 million newtons (3 million pounds)
Second stage	four J-2 engines; total thrust 3.6 million newtons (800 000 pounds)
Third stage	S-IV

Plans for the C-3 were cancelled in favour of a more powerful launch vehicle.

Saturn C-4

First stage	four clustered F-1 engines; total thrust 26.7 million newtons
Second stage	four J-2 engines; total thrust 3.6 million newtons

The C-4 was briefly considered but rejected for the C-5.

Saturn C-5 (renamed Saturn V)**

First stage	S-IC	five F-1 engines, clustered; total thrust 33.4 million newtons
Second stage	S-II	five J-2 engines; total thrust 4.5 million newtons
Third stage	S-IVB	one J-2 engine; total thrust 900 000 N

This was the booster chosen for all the Apollo Moon missions. Its first manned flight was the launch of Apollo 8.

Saturn C-8

First stage	eight F-1 engines; total thrust 53.4 million newtons
Second stage	eight J-2 engines; total thrust 7.1 million newtons
Third stage	one J-2 engine; thrust 889 600 N

Nova

There were several proposals for this rocket. The configuration listed below was typical of the thinking applied to this rocket which was being considered for the direct flight mission profile. Nuclear engines were also considered for the first stage.

First stage	eight F-1 engines; total thrust 53.4 million newtons
Second stage	four liquid-hydrogen M-1 engines; total thrust 21.4 million newtons
Third stage	one J-2 engine; thrust 889 600 N

Appendix 4

Deriving some of the maths

> 'I have just worked out various aspects of the problem of ascending into space with the aid of a reaction machine, rather like a rocket... The scientifically verified mathematical conclusions indicate the feasibility of an ascent into space with the aid of such machines, and, perhaps, the establishment of settlements beyond the confines of the Earth's atmosphere.'

The above was written by Konstantine Tsiolkovsky (1857–1935), a mathematics teacher in the small Russian town of Kaluga. Inspired by the books of Jules Verne, Tsiolkovsky set about precise mathematical calculations to demonstrate some of the basic principles of rocketry that we now accept.

- Liquid fuel rockets would be the best method of designing rockets to explore space and the planets as such engines can be throttled, stopped and restarted. Rockets powered by solid propellant cannot.
- Liquid-fuelled rockets are also more efficient because the hotter and lighter the exhaust gases are, the more efficient is the rocket engine. He also suggested that liquid hydrogen could be burned to achieve hotter and lighter exhaust gases. He was the first person to consider concepts similar to specific impulse.
- He was the first person to calculate the size of the escape velocity from Earth.
- Tsiolkovsky also though up a way of stacking rockets to use the thrust more efficiently. He called this a *sky train*. We now employ this principle in staged rockets (such as the Saturn V) or by discarding 'strap on' boosters (as in the space shuttle).

In chapter 3 I showed how staging is a more efficient way of lifting payloads into orbit based on the rocket equation:

$$\Delta V = u \times \ln\left(1 + \frac{M_{\mathrm{P}}}{M_{\mathrm{R}}}\right)$$

but did not show the proof of the equation. This is done below.

The starting point for studying the motion of any system in which the mass changes must be Newton's second law of motion in the following form:

$$F = \frac{\mathrm{d}}{\mathrm{d}t}(\text{momentum of system}).$$

In chapter 3 we wrote this as

$$F = \frac{\mathrm{d}}{\mathrm{d}t}(mv) = m\frac{\mathrm{d}v}{\mathrm{d}t} + v\frac{\mathrm{d}m}{\mathrm{d}t}$$

but this is not a very useful form for this situation as it assumes that every part of the system (represented by the total mass m) suffers the same change of velocity. In a rocket this is not the case. The propellant mass that is ejected as exhaust suffers a different change in velocity to the remaining rocket and propellant. A much better approach is to consider the momentum change of the elements of the total system.

Let us consider the system to be composed of two parts—the rocket and the propellant to be ejected. If M is the mass of the rocket (including the stored propellant) and Δm the mass of propellant that is to be burnt in the current time interval, Δt, then the change in momentum of the *rocket* is:

$$(M)(v + \Delta v) - Mv = Mv + M\Delta v - Mv = M\Delta v$$

where Δv is the amount by which the velocity of the rocket has changed in time Δt. The momentum of the burnt propellant has also changed in this time. Initially this element of propellant was moving forwards with the rocket at velocity v. At the end of the time interval it is moving forwards with velocity $(v - u)$, where u is the velocity of the propellant relative to the rocket. (If this seems strange, consider that we are viewing this from the ground; if the rocket is moving fast enough then the exhaust exiting backwards relative to the rocket will still be moving forwards relative to the ground.)

Final momentum $-$ initial momentum $= \Delta m(v - u) - \Delta m v = -\Delta m u$ $\therefore$ rate at which the propellant momentum is changing is

$$-u\frac{\Delta m}{\Delta t}$$

which as $\Delta t \to 0$ becomes

$$-u\frac{\mathrm{d}m}{\mathrm{d}t}$$

$\therefore$ total rate of change of momentum of the system (rocket + propellant element) is

$$M\frac{\mathrm{d}v}{\mathrm{d}t} - u\frac{\mathrm{d}m}{\mathrm{d}t}$$

so

$$F = M\frac{\mathrm{d}v}{\mathrm{d}t} - u\frac{\mathrm{d}m}{\mathrm{d}t}$$

where F is the external force applied to the system. In the case of a rocket moving in deep space far from any sources of gravitational pull, the net force acting on the system is zero.

$$\therefore \quad M\frac{\mathrm{d}v}{\mathrm{d}t} = u\frac{\mathrm{d}m}{\mathrm{d}t}.$$

The only way that this equation can be solved is to relate M and m. Fortunately this is quite simple as $\mathrm{d}m/\mathrm{d}t$ is the rate at which the burnt propellant mass is increasing, which must be the same as the rate at which the rocket's mass is *decreasing*.

$$\therefore \quad \frac{\mathrm{d}m}{\mathrm{d}t} = -\frac{\mathrm{d}M}{\mathrm{d}t} \quad \rightarrow \quad M\frac{\mathrm{d}v}{\mathrm{d}t} = -u\frac{\mathrm{d}M}{\mathrm{d}t}$$

so

$$-\int_{M_r+M_f}^{M_r} \frac{\mathrm{d}M}{M} = \frac{1}{u}\int_{v}^{v+\Delta v} \mathrm{d}v$$

and hence

$$-[\ln M]_{M_r+M_f}^{M_r} = \frac{1}{u}(v + \Delta v - v)$$

$$\therefore \quad u \ln\left[\frac{M_r + M_f}{M_r}\right] = \Delta v$$

or

$$\Delta v = u \ln\left[1 + \frac{M_f}{M_r}\right].$$

The situation is quite different if the rocket is in a region where there is a gravitational force. Then the rate of momentum change equation becomes:

$$F = M\frac{\mathrm{d}v}{\mathrm{d}t} - u\frac{\mathrm{d}m}{\mathrm{d}t} = -Mg$$

$$\therefore \quad M\frac{\mathrm{d}v}{\mathrm{d}t} + u\frac{\mathrm{d}M}{\mathrm{d}t} = -Mg$$

or

$$\frac{\mathrm{d}v}{\mathrm{d}t} + \frac{u}{M}\frac{\mathrm{d}M}{\mathrm{d}t} = -g$$

$$\therefore \quad \int_{v}^{v+\Delta v} \mathrm{d}v + u\int_{M_r+M_f}^{M_r} \frac{\mathrm{d}M}{M} = -g\int_{0}^{t} \mathrm{d}t$$

$$\Delta v + u \ln\left[\frac{M_r}{M_r + M_f}\right] = -gt$$

$$\therefore \quad \Delta v = u \ln\left[1 + \frac{M_f}{M_r}\right] - gt.$$

Appendix 5

Further information

Bibliography

Here is a list of books and Internet references that I found useful in compiling the information for this book.

Books

Moon Shot
Alan Shepard and Deke Slayton
Virgin publishing Ltd, 1994

Chariots for Apollo
Courtney G Brooks, James M Grimwood and Loyd S Swenson
Published as NASA Special Publication-4205 in the NASA History Series, 1979
(available on the web)

Moonport: A History of Apollo Launch Facilities and Operations
Charles D Benson and William Barnaby Faherty
Published as NASA Special Publication-4204 in the NASA History Series, 1978

Computers in Space Flight—The NASA Experience
James E Tomayko
http://www.hq.nasa.gov/office/pao/History/computers/Compspace.html

The Anthropic Cosmological Principle
John D Barrow and Frank J Tipler
Oxford University Press, 1986

A Man on the Moon
Andrew Chaikin
Penguin, 1995

The Case for Mars
Robert Zubrin
Touchstone, 1996

All We Did Was Fly to the Moon
Dick Latimer
The Whispering Eagle Press (USA)

What Do You Care What Other People Think?
Richard P Feynman
Unwin Paperbacks, 1988

Coming of Age in the Milky Way
Timothy Ferris
The Bodley Head, 1988

Black Holes and Time Warps—Einstein's Outrageous Legacy
Kipp Thorne
Papermac, 1995

Journal article

Forward R L 1985 Starwisp: an ultra-light interstellar probe *J. Spacecraft and Rockets* vol 22, pp 345–350

Some useful web pages on a variety of related issues

http://www.hq.nasa.gov/office/pao/History/SP-4205/contents.html
Chariots for Apollo

http://www.hq.nasa.gov/office/pao/History/SP-4204/contents.html
Moonport

http://www.hq.nasa.gov/office/pao/History/alsj/frame.html
Apollo lunar surface journal

http://nssdc.gsfc.nasa.gov/planetary/mars/marsprof.html
Information on Mars missions

http://www.grc.nasa.gov/WWW/bpp/
NASA programme on new propulsion systems

http://sunsite.unc.edu/lunar/school/InterStellar/SSD_index.html
A page looking at starship design

http://www.foresight.org/conferences/MNT05/Papers/Bishop/index.html
Another paper looking at novel propulsion ideas

http://www.aleph.se/Trans/Tech/Space/laser.txt
Geoffry Landis' paper on the modified laser propulsion system

http://www.retroweb.com/apollo.html
'Accompanying this personal recollection of the Apollo era are dozens of high-quality Apollo photograph scans, video and audio clips, launch vehicle and crew info/photos, diagrams, a JavaScript lunar-landing simulator, memorabilia, books (with Amazon order links) and links to numerous Apollo and space-related websites'—a good site to visit!

http://users.specdata.com/home/pullo/TWOCOL.HTM
A good page on the lunar module

http://www.farhills.org/s/lees/space/apollo.htm
Space suit information

http://www-sn.jsc.nasa.gov/explore/data/apollo/apollo.htm
Apollo experiments

http://www.jpl.nasa.gov/basics/bsf-toc.htm
Basics of space flight

Where are they now? The crews of Apollo 11 to Apollo 17

Apollo 11

MC – Neil Armstrong
Deputy Associate Administrator for Aeronautics at NASA July 1970–August 1971. Resigned to become Professor of Aeronautical Engineering at the University of Cincinnati. Served on the National Commission on Space from 1985 to 1986 and on the Presidential Commission on the Space Shuttle Challenger Accident in 1986.

CMP – Lt Col Michael Collins (USAF)
Resigned from NASA in January 1970 and was appointed Assistant Secretary of State for Public Affairs. Became Director of the National Air and Space Museum at the Smithsonian Institution in April 1971 and was promoted to Under

Secretary of the Smithsonian in April 1978. Retired from the Air Force with the rank of Major General. Became Vice President, Field Operations, Vought Corporation, Arlington, Virginia, in February 1980. Currently heads Michael Collins Associates, a Washington, DC consulting firm. Has written numerous articles and two books, *Carrying the Fire* and *Liftoff*, as well as a children's book, *Flying to the Moon and Other Strange Places*.

LMP – Col Edwin E 'Buzz' Aldrin (USAF)
Resigned from NASA in July 1971 to become Commandant of the Aerospace Research Pilot's School at Edwards AFB, California. Retired from the Air Force in 1972 and became a consultant for the Comprehensive Care Corporation, Newport Beach, California. Currently resides in southern California and lectures and consults on space sciences with Starcraft Enterprises. Has written several books, including *Return to Earth* and *Men From Earth*.

Apollo 12

MC – Comdr Charles Conrad (Navy)
Retired the Navy, with the rank of Captain, and from NASA on February 1, 1974. Became Vice President and Chief Operating Officer of the American Television and Communications Corporation. Currently Vice President for International Business Development, McDonnell Douglas Space Systems Co.

Charles 'Pete' Conrad died in a motorcycle accident on 9 July 1999.

CMP – Comdr Richard F Gordon (Navy)
Retired from NASA and the Navy, with the rank of Captain, on 1 January 1972, to become Executive Vice-President of the New Orleans Saints football team, a position he resigned on 1 April 1977. Worked as Vice President, Operations, at Scott Science and Technology, Inc., and is currently Chairman, Astro Science Corporation in Los Angeles, and President, Space Age America, Inc.

LMP – Comdr Alan L Bean (Navy)
Retired from the Navy with the rank of Captain in October 1975 and from NASA on 26 June 1981, to work as a space artist.

Apollo 13

MC – Capt James A Lovell (Navy)
Became Deputy Director of Science and Applications at the Johnson Space Center in May 1971. Left NASA and the Navy in March 1973 to become President and Chief Executive Officer of the Bay-Houston Towing Company,

but resigned that position on 1 January 1977, to become President of Fisk Telephone Systems in Houston. On 1 January 1981, he became Group Vice President of Centel Corporation, and was later promoted to Executive Vice President. Currently, President, Lovell Communications.

CMP – John L Swigert
In 1982 he was elected to the US House of Representatives. However, he died from complications due to cancer the week before the swearing-in ceremony.

LMP – Fred W Haise
Commanded one of the two crews that flew shuttle approach and landing tests. Resigned from NASA on 29 June 1979 and is currently President, Technical Services Division, Grumman Corp.

Apollo 14

MC – Capt Alan B Shepard (Navy)
Was chosen with the first group of astronauts in 1959. Was the pilot of Freedom 7 and thus the first American in space. Was subsequently grounded due to an inner ear ailment until 7 May 1969 (during which time he was chief of the Astronaut Office). Commanded Apollo 14, and in June 1971 resumed duties as chief of the Astronaut Office. Retired from NASA and the Navy on 1 August 1974, with the rank of Rear Admiral, to join the Marathon Construction Company of Houston, Texas, as partner and chairman. Later, President, Seven/Fourteen Enterprises, Houston.

Alan Shepard died on 21 July 1998 after a long battle with leukaemia.

CMP – Maj Stuart A Roosa (USAF)
Retired from the Air Force with the rank of Colonel. Resigned from NASA on 1 February 1976 to become Vice President for International Affairs, US Industries Middle East Development Company, based in Athens, Greece. Returned to the United States in 1977 to become President of Jet Industries in Austin, Texas. Later worked in private business in Austin and was owner and President, Gulf Coast Coors, Inc., Gulfport, Mississippi.

Stu Roosa died on 26 December 1994 of complications from pancreatitis.

LMP – Comdr Edgar D Mitchell (Navy)
Retired from the Navy, with the rank of Captain, and from NASA in October 1972. Founded the Institute of Noetic Sciences in Palo Alto. Was President, Edgar Mitchell Corporation (EMCO), from 1974 until 1978. Later became Chairman of Mitchell Communications Company in Florida. Wrote a book entitled *Psychic Exploration: A Challenge for Science*.

Apollo 15

MC – Col David R Scott (USAF)
Became Special Assistant for Mission Operations for the Apollo–Soyuz Test Project in July 1972, and was appointed Director, NASA Dryden Flight Research Center, in April 1975. Resigned 30 October 1977, to found his own company, now Scott Science and Technology, Inc. In 1994/1995, he served as the main technical consultant on the film *Apollo 13* and, in 1997, served in the same capacity on the Tom Hanks/HBO series, *From the Earth to the Moon*.

CMP – Maj Alfred M Worden (USAF)
Assigned to NASA Ames Research Center as Director of Advanced Research and Technology in September 1972. Retired from NASA and the Air Force on 1 September 1975, and worked as President of the Energy Management Consulting Company and later as President of M W Aerospace Inc. Currently Staff Vice President, B F Goodrich Company, Aerospace Division. Wrote a book of poetry entitled, *Hello Earth: Greetings from Endeavour*, and a children's book, *A Flight to the Moon*.

LMP – Lt Col James B Irwin (USAF)
Irwin retired from NASA and the Air Force in July 1972 to form and lead a religious organization, High Flight Foundation, in Colorado Springs, Colorado.

James Irwin died of a heart attack on 8 August 1991.

Apollo 16

MC – Capt John W Young (Navy)
Young became the Chief of the Astronaut Office in 1975 but remained active in development and testing of the shuttle. Commander of the first shuttle flight in April 1981. Commander of STS-9. Young is still a member of the astronaut corps, although not on active flight status.

CMP – Lt Comdr Thomas K Mattingley (Navy)
Head of the astronaut support team for the shuttle programme from 1973 until 1978. Mattingley was Commander of the fourth mission (STS 4) in June 1982. January 1985, commanded the 15th mission, STS-51C. After leaving NASA, he served as Director, Space Sensor Systems, US Navy Space and Naval Warfare Systems Command. Now Director, Utilization and Operation, Grumman Space Station Program Support Division, Reston, Virginia.

LMP – Lt Col Charles M Duke (USAF)
Duke retired from the Astronaut Corps in 1975 and is active in private business and as a Christian lay witness.

Apollo 17

MC – Capt Eugene A Cernan (Navy)
Cernan helped in the planning for Apollo–Soyuz and acted for the programme manager as the senior US representative in discussions with the USSR. He retired from NASA and the Navy, with the rank of Captain, on 1 July 1976 to become a consultant in the energy and aerospace businesses and a television commentator.

CMP – Comdr Ronald E Evans (Navy)
Evans served as backup command module pilot for the 1975 Apollo–Soyuz joint flight and then transferred to the shuttle program. Retired from NASA in 1977. He worked as Executive Vice-President of Western America Energy Corporation in Scottsdale, Arizona until 1978, then as Manager, Space Systems Marketing for Sperry Flight Systems, Phoenix, Arizona, and later as a marketing consultant.

Ronald Evans died of a heart attack on 7 April 1990.

LMP – Dr Harrison H Schmitt
After Apollo 17, he played an active role in documenting the Apollo geologic results and also organized NASA's Energy Program Office. In August 1975, Schmitt resigned from NASA to seek election as a United States Senator for New Mexico. He served one term and, notably, was the ranking Republican member of the Science, Technology, and Space Subcommittee. He was defeated in a re-election bid in 1982. Now a consultant in business, geology, space, and public policy.

NASA centres

NASA headquarters, Washington, DC
Admin and budget centre

Ames Research Center, Moffet Field, CA
Specializes in research geared toward creating new knowledge and new technologies that span the spectrum of NASA interests

Dryden Flight Research Center, Edwards AFB, CA
Civil aeronautical flight research; development and operations of the space shuttle; developing piloted and uninhabited aircraft test beds for research and science missions

Goddard Institute for Space Studies, New York, NY
Division of Goddard Space Flight Center

Goddard Space Flight Center, Greenbelt, MD
Research into Earth and environment using observations from space

Independent Validation & Verification Facility, Fairmont, WV
Software research and validation

Jet Propulsion Laboratory, Pasadena, CA
Leading US centre for robotic exploration of the solar system

Johnson Space Center, Houston, TX
Design and testing of manned space flight systems, selection and training of astronauts, mission planning and control, medical and scientific experiments in flight

Kennedy Space Center, Florida
Primary launch facility for manned missions

Langley Research Center, Hampton, VA
Aeronautics, atmospheric sciences and space technology

Lewis Research Center, Cleveland, OH
New propulsion, power, and communications technologies

Marshall Space Flight Center, Huntsville, AL
Developing space transportation and propulsion systems and for conducting microgravity research and space optics manufacturing technologies

Moffett Federal Airfield, Mountain View, CA
A research and development facility for active and reserve military units, government agencies and companies with links to NASA providing airfield operations

Stennis Space Center, Mississippi
Centre for testing and flight certifying rocket propulsion systems for the space shuttle and future generations of space vehicles

Wallops Flight Facility, Wallops Island, VA
Operational test site for the next generation of low-cost launch technologies linked to Goddard

White Sands Test Facility, White Sands, NM
Testing and evaluation of spacecraft materials, components, and propulsion systems to enable the safe human exploration and utilization of space.

Index

Page numbers in italics are references to chapter endnotes, those in bold are figure captions or figures showing the item concerned.

B

C

G

H

I

J

K

L

M

N

T

V

W

X

Z